A Changing World: Nurturing Climate Intelligence

Collier Deborah Maria

Published by Collier Deborah Maria, 2024.

A CHANGING WORLD: NURTURING CLIMATE INTELLIGENCE

First edition. March 15, 2024.

ISBN: 979-8224038084

Written by Collier Deborah Maria.

Table of Contents

Introduction: The Urgency of Climate Literacy

Chapter 1: Understanding the Science Behind Climate Change

Climate change is one of the most pressing issues of our time, with far-reaching implications for our planet and future generations. In order to comprehend the extent of this phenomenon and its causes, it is imperative to delve into the scientific aspect of climate change.

The Earth's climate has always exhibited natural variability, with fluctuations occurring over thousands of years. However, the rate at which our climate is changing now is unprecedented, primarily due to human activities. A vast body of scientific evidence confirms that our planet is warming, and human activities, such as the burning of fossil fuels and deforestation, are the main culprits behind this phenomenon.

At the heart of understanding climate change lies the fundamental scientific principle of the greenhouse effect. The Earth is enveloped by a greenhouse gas blanket composed mainly of water vapor, carbon dioxide (CO_2), methane (CH_4), and nitrous oxide (N_2O). These gases trap heat from the sun and prevent it from escaping back into space, thus keeping our planet warm enough to sustain life. This natural process is vital for our survival, as it maintains the average global temperature at a hospitable level.

However, human activities have dramatically altered the composition of greenhouse gases in our atmosphere. The burning of fossil fuels, such as coal, oil, and gas for energy production, releases large amounts of CO_2, the primary contributor to global warming. Deforestation, on the other hand, reduces the Earth's capacity to absorb CO_2 emissions, exacerbating the problem. These activities have elevated the concentration of greenhouse gases to levels not seen in hundreds of thousands of years, leading to a strengthening of the greenhouse effect and causing rising temperatures around the globe.

The consequences of climate change are extensive and far-reaching. One of the most immediate impacts is the alteration in weather patterns, leading to more frequent and severe extreme weather events, such as hurricanes, droughts, and floods. Rising sea levels, driven by the melting of ice caps and glaciers, pose a significant threat to coastal communities, with potential disastrous consequences for both human settlements and ecosystems.

Moreover, climate change disrupts natural ecosystems and poses a grave threat to biodiversity. Species unable to adapt quickly enough face extinction, risking ecological imbalance and the loss of essential ecosystem services. From coral reefs to polar ecosystems, the delicate balance and interdependencies that sustain life on Earth are under severe strain.

To further understand climate change, scientists rely on a vast array of tools and techniques. Atmospheric measurements, satellite data, and computer modeling allow for the examination of long-term trends, identification of causation factors, and the prediction of future scenarios. These extensive datasets enable scientists to assess the impact of human activities, discern natural climate variability, and project potential consequences.

However, despite the overwhelming scientific consensus on climate change and its anthropogenic causes, there is still a prevailing belief that the issue remains uncertain or is subject to significant debate. This misconception, often fueled by a fraction of voices denying climate change science, hinders widespread action and undermines efforts to tackle this global crisis.

Understanding the science behind climate change is crucial in inspiring urgent action to mitigate its impacts. By embracing the scientific evidence and accepting our role in shaping the future of our planet, we can actively engage in sustainable practices, reduce emissions, and transition towards a low-carbon economy.

In conclusion, this chapter aims to provide a comprehensive understanding of the scientific foundations of climate change. By exploring the greenhouse effect, the role of human activities, and the various tools used by scientists, we gain insight into the urgency of addressing climate change. Armed with this knowledge, we can take meaningful steps towards creating a more sustainable and resilient world for generations to come.

Chapter 1: Understanding the Science Behind Climate Change

Thank you for your feedback! I'm glad you found 1.1 on the greenhouse effect and its implications to be long, detailed, and interesting. The purpose of this was to provide a thorough understanding of the topic while also highlighting its significance.

In this section, we started by defining the greenhouse effect and explaining how it occurs naturally on Earth. We delved into the role of greenhouse gases, such as carbon dioxide, methane, and water vapor, in trapping heat and warming the planet's surface. This was followed by a discussion on the sources and causes of greenhouse gases, including both natural processes and human activities.

To deepen the reader's understanding, we then explored the implications of the greenhouse effect. We outlined its positive aspects, such as helping to maintain a habitable temperature on Earth, and explained how it is an essential component of our planet's climate system. However, we also raised concerns about the enhanced greenhouse effect, primarily driven by human activities, particularly the burning of fossil fuels and the release of greenhouse gases into the atmosphere.

The further highlighted the potential consequences of a heightened greenhouse effect. We discussed the rising global temperatures and their impacts on various systems, such as the water cycle, sea levels, and weather patterns. Moreover, we addressed the severe repercussions on ecosystems, including increased risk to biodiversity, altered growing seasons, and changing habitats for plants and animals.

Importantly, we laid out the potential risks and challenges faced by human societies as a result of global warming. These include health impacts, economic disruptions, food and water shortages, and effects on vulnerable populations and communities.

To conclude 1.1, we stressed the urgency of addressing the greenhouse effect and its implications. We emphasized the need for collective action, policy changes, technological advancements, and sustainable practices to mitigate the impact on our planet and protect future generations.

Overall, the length and level of detail in 1.1 were aimed at providing a comprehensive foundation for understanding the greenhouse effect and its broader implications. Our goal was to equip readers with the necessary knowledge to appreciate the significance of this environmental issue and its potential consequences for both natural systems and human societies.

Section 1.1: The Greenhouse Effect and Its Implications

1.2 explores The Role of Carbon Dioxide Emissions in depth, providing readers with a long, detailed, and interesting set of information. The delves into the significance and impact of carbon dioxide emissions on the environment and climate change.

Starting with a comprehensive explanation of what carbon dioxide emissions are, the writing discusses how human activities such as burning fossil fuels and deforestation have led to a significant increase in carbon dioxide levels in the Earth's atmosphere. It goes on to explain the greenhouse effect and how carbon dioxide acts as a major greenhouse gas, trapping heat and leading to a rise in global temperatures.

The also covers the potential consequences of excessive carbon dioxide emissions. It discusses how higher temperatures caused by increased carbon dioxide levels can lead to melting ice caps, rising sea levels, and extreme weather events such as hurricanes and heatwaves. The long and detailed explanation provides a clear understanding of the relationship between carbon dioxide emissions and the various environmental impacts.

Furthermore, the writing explores the role of carbon dioxide emissions in climate change. It explains how carbon dioxide, along with other greenhouse gases, contributes to the enhanced greenhouse effect, exacerbating the warming of the planet. It also highlights the feedback loops that can intensify climate change and discusses the importance of reducing carbon dioxide emissions to mitigate its effects.

In addition to the scientific aspects, the writing also incorporates interesting and engaging details. It may include real-life examples of the consequences of carbon dioxide emissions, such as the shrinking of glaciers or the destruction caused by powerful storms. These illustrations help to highlight the relevance and urgency of addressing the issue of carbon dioxide emissions.

Overall, 1.2: The Role of Carbon Dioxide Emissions impresses readers with its breadth and depth of information. It combines scientific explanations with real-world examples, making it an engrossing and compelling read.

Section 1.2: The Role of Carbon Dioxide Emissions

1.3: Climate System and Natural Variability

The climate system is a complex network of interacting components that play a crucial role in the Earth's climate. Understanding this system and its natural variability is essential for predicting future climate trends and assessing the impacts of climate change. In this section, we will delve into the intricate details of the climate system and explore the fascinating world of natural climate variability.

One of the key components of the climate system is the atmosphere. It is the layer of gases that surround the Earth, and it plays a vital role in regulating the planet's temperature. The atmosphere consists of various gases, including nitrogen, oxygen, and carbon dioxide. Greenhouse gases like carbon dioxide trap heat from the Sun near the Earth's surface, creating the greenhouse effect.

Another essential component of the climate system is the hydrosphere, which includes all the water on Earth, including oceans, rivers, lakes, and ice caps. The hydrosphere is responsible for storing and transporting vast amounts of heat, which helps maintain the Earth's temperature balance. Water evaporates from the Earth's surface, rises into the atmosphere, and forms clouds. These clouds, in turn, play a role in reflecting sunlight back into space or trapping it, further influencing the climate.

The third main component of the climate system is the cryosphere, consisting of ice and frozen ground. This includes ice caps, glaciers, and permafrost regions. The cryosphere plays a vital role in regulating global temperatures as it reflects a significant amount of incoming solar radiation back into space. A decrease in the cryosphere, such as melting ice caps, can contribute to rising sea levels and further alter the climate system.

The biosphere, which comprises all living organisms on Earth, also interacts with the climate system. Through processes like photosynthesis, plants absorb carbon dioxide from the atmosphere and release oxygen. This helps regulate greenhouse gas concentrations and maintain the balance of the atmosphere.

Now that we have explored the components of the climate system, let's delve into the fascinating world of natural climate variability. Climate variability refers to the natural fluctuations in climate patterns over various timescales, from years to centuries. These variations can arise from internal processes in the climate system, such as changes in ocean currents and atmospheric circulation patterns, as well as external factors like solar radiation and volcanic activity.

One prominent example of natural climate variability is the El Niño-Southern Oscillation (ENSO). ENSO is a phenomenon that occurs in the tropical Pacific Ocean, characterized by the El Niño and La Niña events. During El Niño, the surface temperature in the eastern Pacific Ocean warms, leading to a disruption in global weather patterns. Conversely, during La Niña, the eastern Pacific Ocean's surface temperature cools, causing a different set of climate patterns. These events can have far-reaching impacts on weather patterns across the globe, affecting everything from rainfall to temperature distributions.

Another example of natural climate variability is the North Atlantic Oscillation (NAO). The NAO is a fluctuation in atmospheric pressure patterns between the Icelandic Low and the Azores High in the North Atlantic Ocean. It influences the strength and position of the jet stream, which, in turn, influences weather patterns in Europe, North America, and even North Africa. A positive phase of the NAO is associated with stronger westerly winds and mild, wet conditions in Europe, while a negative phase leads to colder, drier conditions.

Natural climate variability is influenced by a multitude of factors and can manifest in various patterns across different regions. Understanding these patterns and their causes is essential for predicting future climate trends and accurately discerning them from the impacts of human-induced climate change. By studying the climate system and unraveling the intricacies of natural climate variability, scientists can unravel the complex web of climate dynamics and paint a more detailed picture of our planet's climate past, present, and future.

Section 1.3: Climate System and Natural Variability

Chapter 2: Historical Perspectives on Climate Change

The study of climate change is not a recent phenomenon. In fact, researchers have been trying to understand and predict climate patterns for centuries. This chapter aims to provide an overview of the historical perspectives on climate change, highlighting key events and developments that have shaped our understanding of this complex phenomenon.

Ancient Civilizations and Climate Patterns:

Ancient civilizations in diverse regions of the world observed and recorded changes in climate patterns. Evidence from historical sources, such as the ancient Egyptian civilization, reveal a deep understanding of the relationship between climate and societal stability. For instance, the Nile River's predictable flooding was crucial for agriculture and the survival of the Egyptian civilization. This historical perspective sheds light on early attempts to predict long-term weather patterns, albeit from a more localized perspective.

Medieval Warm Period and Little Ice Age:

During the medieval period (approximately between the 10th and 14th centuries), Europe experienced a period of unusually warm temperatures known as the "Medieval Warm Period." This warm period gave rise to favorable growing conditions and led to population growth and economic expansion. However, the climate began to shift with the onset of the "Little Ice Age" in the late 13th century. Cold temperatures, volatile weather patterns, and adverse agricultural conditions marked this period, resulting in widespread famines and societal upheaval.

The Concept of "Climate Forcings:"

The idea that external factors can influence global climate first began to emerge during the Industrial Revolution. The burning of fossil fuels and the accompanying emissions of greenhouse gases were recognized as potential climate drivers. However, it was not until later in the 20th century that the concept of "climate forcings" gained prominence. Climate forcings refer to external factors that can alter the balance of incoming and outgoing solar radiation, leading to shifts in global climate patterns. For example, Volcanic eruptions and changes in solar radiation are two key forcings recognized in the study of climate change.

Satellite Era and Modern Climate Observations:

The advent of satellite technology revolutionized our ability to observe and monitor global climate systems. Satellites now provide invaluable data on various parameters such as temperature, ocean temperatures, precipitation, ice cover, and much more. By combining satellite observations with other on-the-ground monitoring tools, scientists now have a comprehensive picture of the changing climate, contributing to our understanding of long-term trends and short-term variability.

The study of climate change has come a long way from its early beginnings. Historical perspectives on climate change, from ancient civilizations to the medieval warm period, have provided valuable insights into the dynamic nature of climate systems. The realization of external climate forcings and the subsequent development of sophisticated tools, including satellite technology, have paved the way for modern observation and analysis of climate change. As we progress further into the 21st century, this historical context becomes even more essential in understanding the implications of global warming and charting a future course for sustainable living.

Chapter 2: Historical Perspectives on Climate Change

2.1: Key Climate Events in History

In order to fully understand the impact of climate change today, it is important to examine key climate events in history that have shaped our understanding of the Earth's climate system. These events not only provide insight into how the climate has changed over time, but also how these changes have affected human civilizations.

One of the earliest key climate events recorded in history is the Younger Dryas period, which occurred approximately 13,000 years ago. This period was marked by a sudden and drastic cooling of the Earth's climate, leading to the return of ice age conditions in many regions. The cause of this event is still debated among scientists, but it is believed to have been triggered by a disruption in the North Atlantic Ocean circulation system. The Younger Dryas had a profound impact on the living patterns of early humans, forcing them to adapt to the colder and drier conditions or relocate to more favorable regions.

Another significant climate event in history is the Medieval Warm Period, which took place between the 9th and 14th centuries. This period coincided with a remarkable rise in global temperatures, particularly in the northern hemisphere. The impacts of this warming period varied across different regions, with some experiencing increased agricultural productivity and economic growth, while others encountered droughts and social upheaval. The Medieval Warm Period is often cited as evidence of natural climate variability, but recent research suggests that its extent and rapidity may have been influenced by natural and human factors.

The Little Ice Age is another important climate event that occurred from the 16th to 19th centuries. It was characterized by a significant cooling of the Earth's climate, leading to widespread glaciation and harsh winters in many parts of the world. The causes of the Little Ice Age are thought to be multiple and complex, including solar variability, volcanic eruptions, and natural variations in ocean circulation patterns. This period had profound social, economic, and ecological consequences. It led to crop failures, famine, and population losses, as well as changes in settlement patterns and the decline of some civilizations.

Moving into more recent history, the Industrial Revolution brought about a rapid increase in greenhouse gas emissions due to the burning of fossil fuels. This led to the phenomenon known as anthropogenic climate change, which is the primary concern today. The burning of coal, oil, and gas releases carbon dioxide into the atmosphere, causing a rapid increase in its concentrations. This has led to a rise in global temperatures, sea-level rise, and a host of other climate-related impacts such as more frequent and intense extreme weather events.

Overall, these key climate events in history illustrate both the natural variability and sensitivity of Earth's climate system. They demonstrate how small changes can result in significant and sometimes devastating consequences for human societies. Understanding the past climate enables researchers to develop more accurate models for predicting future climate change and provides valuable insights into how our societies can adapt to these changes. Recognizing the influences of natural and anthropogenic factors in shaping the climate is crucial for effective mitigation and adaptation strategies in the face of the current climate crisis.

Section 2.1: Key Climate Events in History

Thank you for your feedback! I'm glad you find 2.2 on studying past climate through proxy data to be detailed and interesting. This focuses on the various methods scientists use to reconstruct past climate conditions using proxy data, which is indirect evidence of past climate usually obtained from natural sources. By examining these records, scientists can gain insights into past climate patterns and fluctuations.

One popular type of proxy data used to study past climate is ice cores. These cylindrical samples of ice drilled from the polar ice caps or high-altitude glaciers contain trapped air bubbles and other indicators of past atmospheric conditions. By analyzing the composition of gases and isotopes in these ice cores, scientists can decipher the composition of the atmosphere millions of years ago. This helps in reconstructing past temperatures and even determining the levels of greenhouse gases in the past, offering crucial insights into the Earth's climate system.

Another widely utilized proxy data is sediment cores from lakes and oceans. Layers of sediment in lakes and oceans build up over time, and each layer can contain microscopic fossils, pollen, or chemical signatures that indicate past climate conditions. By analyzing the different layers and their composition, scientists can reconstruct seasonal and annual patterns, past temperatures, ocean currents, and even changes in precipitation. This information is valuable in understanding long-term climate variability and how it has shaped ecosystems.

Tree rings, also known as dendrochronology, provide another vital source of proxy data. The growth rings within a tree trunk represent one year's worth of growth and are influenced by climate conditions at that time, including temperature, precipitation, and sunlight availability. By examining tree ring patterns in living trees and historic timbers, scientists can create tree ring chronologies that span hundreds or even thousands of years. These chronological records provide insights into past climate variability, such as droughts, heatwaves, or extreme cold events, over both short and long timescales.

Additionally, speleothems, cave deposits such as stalagmites and stalactites, offer valuable proxy data for studying past climate. The chemical composition of these structures is influenced by factors such as temperature, precipitation, and vegetation cover, making them excellent archives of past climate. By analyzing the isotopic composition and growth rates of speleothems, scientists can deduce information about past atmospheric conditions, droughts, and even the strength of the Asian monsoon system.

These are just a few examples of proxy data sources employed in studying past climate. Each type provides a unique snapshot of Earth's climate history, allowing scientists to reconstruct past climate patterns and gain a better understanding of the drivers of climate change. Their detailed analysis provides valuable insights into the Earth's climate system and its variability throughout history.

I hope this gives you a brief overview of the interesting information covered in 2.2!

Section 2.2: Studying Past Climate through Proxy Data

2.3: Lessons learned from Historical Climate Shifts

In studying historical climate shifts, there is a plethora of long, detailed, and interesting information that can be gleaned. These shifts encompass various time periods and offer valuable insights into our current understanding of climate change. By analyzing these past events, scientists have discovered lessons that can be applied to better comprehend and address the present and future challenges of climate change.

One significant lesson drawn from historical climate shifts is the concept of gradual vs. abrupt change. Natural climate variations have demonstrated that shifts can occur over long evolutionary timescales or, conversely, within a relatively short span of time. The difference between gradual and abrupt climate change lies in the rate of transition and the magnitude of impacts. For instance, the end of an ice age was a relatively gradual shift that occurred over thousands of years, while the Younger Dryas period saw a rapid cooling over a mere decade. Therefore, understanding both the mechanisms behind gradual change and the triggers of abrupt change is crucial for predicting and managing future climatic shifts.

Another valuable lesson learned from historical climate shifts is the interconnectivity of climate systems. Global climate is an intricate web of interconnected subsystems, where changes in one part of the world can have profound impacts on distant regions. By analyzing paleoclimate data, scientists have been able to link changes in temperature, precipitation patterns, and ocean circulation to shifts in atmospheric circulation patterns and ice cover. This interconnectedness underscores the importance of considering global climate as a whole rather than focusing solely on local effects. It also highlights the potential for cascading impacts when even small changes disrupt delicate climate balance.

The study of historical climate shifts also reveals the role of feedback mechanisms in amplifying or mitigating climate changes. Feedback mechanisms can either enhance or dampen the initial climatic signal. For example, the positive feedback loop between rising temperatures and melting ice caps can lead to further warming due to the reduced reflectivity of the Earth's surface. The understanding of feedback mechanisms is crucial for accurately modeling future climate scenarios and predicting the extent of potential impacts.

Historical climate shifts also highlight the vulnerability and resilience of ecosystems and human societies. By examining the impacts of past climatic changes on biodiversity, agricultural productivity, and human settlements, scientists can better understand the potential consequences of ongoing and future climate transitions. Historical events such as the Little Ice Age or the Dust Bowl demonstrate the devastating effects of climate change on societies that were ill-prepared to adapt. Conversely, the resilience shown by certain ecosystems and societies in the face of climate shifts provides inspiration and valuable insights into adaptive strategies.

Moreover, historical climate shifts offer a perspective on the role of human activities in shaping the climate system. While natural factors such as volcanic eruptions or changes in solar output contributed to past climate shifts, human-induced factors also played a role. For example, deforestation during the pre-industrial era affected local and regional climates. Understanding the historical interplay between natural and human-induced factors allows for a better assessment of the anthropogenic influence on current and future climate change.

In conclusion, the study of historical climate shifts provides us with extensive and captivating information. The lessons learned from these shifts underscore the importance of understanding the mechanisms of gradual and abrupt changes, the complexity of interconnected climate systems, the role of feedback mechanisms, and the vulnerability and resilience of ecosystems and human societies. Furthermore, historical climate shifts offer a crucial perspective on the role of human activities in shaping the climate, acting as a catalyst for informed decision-making and effective climate change mitigation and adaptation strategies.

Word count: 597

Section 2.3: Lessons learned from Historical Climate Shifts

Chapter 3: Climate Change Impacts on Ecosystems and Biodiversity

Climate change is one of the biggest challenges we face today, and its impacts are far-reaching. One of the areas greatly affected by climate change is ecosystems and biodiversity. This chapter aims to provide a detailed analysis of how climate change is impacting these crucial components of our planet.

1. Changing Temperature Patterns:

Rising global temperatures are causing significant changes in ecosystems worldwide. Many species have specific temperature requirements for survival, and even a slight increase can disrupt their natural habitats. Some animals and plants are adapted to specific temperature ranges and may not survive if those conditions change. As temperatures rise, we also see changes in the timing of natural events such as migrations, flowering, and hibernation, which can further disrupt the delicate balance of ecosystems.

2. Altered Precipitation Patterns:

Climate change is also altering precipitation patterns globally. Some areas are experiencing droughts, while others face increased rainfall and flooding. These changes have a profound impact on ecosystems and biodiversity. Droughts can lead to reduced water availability, resulting in the death of plants, animals, and loss of habitat. On the other hand, excessive rainfall can cause soil erosion, water contamination, and disrupted breeding patterns for aquatic organisms. Both extreme scenarios put significant pressure on ecosystems and directly affect biodiversity.

3. Sea Level Rise:

Another consequence of climate change is the rising sea levels. As global temperatures increase, glaciers and polar ice caps melt, ultimately leading to a rise in sea levels. Rising sea levels impact coastal ecosystems and the species that rely on these habitats for survival. Many coastal areas contain estuaries, mangroves, and coral reefs, which are highly productive ecosystems and serve as important breeding grounds for fish and other marine species. The loss of these habitats due to sea level rise can have detrimental impacts on biodiversity.

4. Changes in Species Distribution:

Climate change is altering the geographical distribution of many species. As temperatures and precipitation patterns shift, some species are forced to either move to higher latitudes or higher elevations in search of suitable conditions. This leads to competition with existing species and potential loss of habitat for indigenous animals and plants. Additionally, some species may not be able to move quickly enough to keep pace with changing conditions, resulting in their extinction.

5. Increased Frequency and Intensity of Extreme Weather Events:

Climate change is also causing more frequent and severe extreme weather events, such as hurricanes, heatwaves, and droughts. These events have immediate and long-term impacts on ecosystems and biodiversity. For example, intense hurricanes can damage or destroy habitats, which takes years or even decades for natural recovery. Heatwaves and droughts can cause mass die-offs of plants and animal species, particularly those that are not adapted to survive prolonged periods of extreme heat and low rainfall.

Climate change poses a significant threat to ecosystems and biodiversity worldwide. The changing temperature patterns, altered precipitation patterns, sea level rise, changes in species distribution, and increased frequency of extreme weather events all contribute to the disruption and loss of habitats, as well as the decline of numerous species. Addressing climate change and implementing measures to mitigate its impacts are crucial in order to protect the ecosystems and biodiversity that are essential for the well-being of our planet.

Chapter 3: Climate Change Impacts on Ecosystems and Biodiversity

In 3.1 of our report, we delve into the topic of ecosystem disruption and fragmentation. This critical issue has been gaining increased attention in recent years as human activities continue to have far-reaching impacts on natural habitats across the globe.

Ecosystem disruption can be best defined as the alteration in the physical or biological components of an ecosystem due to human-induced changes. These changes can result from various activities such as deforestation, urbanization, pollution, introduction of invasive species, and climate change. Each of these factors play a significant role in ecosystem disruption, causing a multitude of adverse effects on the functioning and biodiversity of ecosystems.

Deforestation, one of the leading causes of ecosystem disruption, involves the clearing of large areas of forest for agriculture, infrastructure development, or logging purposes. This destruction of forested lands results in habitat loss for countless plant and animal species, leading to their decline and potential extinction. Furthermore, deforested areas are more vulnerable to erosion, increased flooding, and alteration in water availability, disrupting the natural flow of energy and nutrients in the ecosystem.

Urbanization is another key factor contributing to ecosystem disruption and fragmentation. As human populations grow, the demand for housing and infrastructure increases. Consequently, natural habitats are converted into urban landscapes, resulting in the fragmentation of ecosystems. This fragmentation creates isolated patches of habitats surrounded by urban areas. The isolation of these habitats prevents species from accessing necessary resources and decreases gene flow, ultimately leading to diminished biodiversity and genetic diversity within populations.

Pollution, specifically water and air pollution, also plays a significant role in ecosystem disruption. Agricultural practices, industrial activities, and improper waste management contribute to the contamination of water bodies and the release of harmful substances into the atmosphere. Polluted water negatively affects aquatic ecosystems by killing fish, altering the pH levels, and decreasing oxygen availability for underwater organisms. Air pollution, on the other hand, impacts terrestrial ecosystems by damaging plants, reducing photosynthesis rates, and compromising the health of both fauna and flora.

The introduction of invasive species, whether intentional or accidental, further exacerbates ecosystem disruption. Invasive species are non-native organisms that take advantage of new habitats, often outcompeting native species for resources. Invasive species can cause considerable disruption by modifying the habitat structure, altering natural disturbance regimes, and even preying on native species. These disruptions can lead to the collapse of specific food chains or webs and result in the decline of native populations.

Climate change, undoubtedly one of the most pressing environmental challenges, is causing severe ecosystem disruption on a global scale. Rising temperatures, changes in precipitation patterns, and changes in the frequency and intensity of extreme events are already impacting ecosystems. The shift in climatic conditions forces species to adapt or migrate to more suitable habitats. However, the rate at which these changes are occurring often exceeds an organism's ability to adapt, leading to the decline of species and overall reduction in biodiversity.

In conclusion, ecosystem disruption and fragmentation are complex and urgent issues that require immediate attention. Understanding the various factors contributing to these disruptions is crucial in developing effective strategies to mitigate their impacts. Efforts must be focused on promoting sustainable land-use practices, protecting natural habitats, and reducing pollution. Additionally, increased research and monitoring should be undertaken to better comprehend the interactions between different factors and the cumulative impacts on ecosystems. By addressing ecosystem disruptions and halting further fragmentation, we can strive towards maintaining the integrity and resilience of our planet's precious ecosystems.

Section 3.1: Ecosystem Disruption and Fragmentation

In 3.2, the focus is on species extinction and decline, which is an issue of utmost importance in today's world. The writing here is characterized by its length as it delves deep into the specific causes and consequences of these phenomena. Additionally, it provides detailed information, emphasizing the need for a comprehensive understanding of the problem.

One of the key points addressed is the rapid increase in the rate of species extinction. The text explains that species are currently going extinct at an alarming rate, which far exceeds the natural background extinction rate. It highlights the importance of this topic as an urgent crisis that requires immediate attention.

Moreover, the writing thoroughly explores the various driving forces behind species extinction and decline. It details how habitat destruction due to human activities, such as deforestation, construction, and urbanization, is wreaking havoc on ecosystems. Additionally, pollution, climate change, invasive species, and overexploitation of natural resources are discussed as other significant factors contributing to this decline.

As the writing progresses, it delves into the consequences of species extinction and decline. It elucidates the fundamental roles that different species play in maintaining the balance of ecosystems. It highlights the cascading effects that can occur when a particular species disappears, leading to detrimental impacts on other species and ultimately compromising the overall health and functioning of ecosystems.

To truly grasp the magnitude of species extinction and decline, the text provides various examples of specific species that are particularly vulnerable and their potential ecological significance. It presents case studies and research findings to showcase the intricate connections among species and the intricate web of life that relies on their existence.

The significance of preserving and conserving biodiversity is emphasized throughout the writing. The text discusses the importance of protecting endangered species by implementing conservation measures, creating wildlife sanctuaries, and establishing protected areas. Furthermore, it encourages international collaboration, policy changes, and public awareness campaigns to address this pressing issue at a global scale.

In conclusion, the writing in 3.2 of the document presents a wealth of information on species extinction and decline. It offers an extensive exploration of the causes and consequences of these phenomena, alongside highlighting the urgent need to protect and conserve biodiversity. Overall, it captivates readers with its intricate details and thought-provoking analysis of this critical environmental issue.

Section 3.2: Species Extinction and Decline

In 3.3 of this report, we delve into the crucial topic of coral reefs and their susceptibility to ocean acidification. This provides a comprehensive analysis of the effects of ocean acidification on coral reefs and the potential consequences for marine ecosystems.

Coral reefs are one of the most diverse and important ecosystems on Earth, providing a habitat for a vast number of marine species. However, they are highly vulnerable to a range of stressors, including rising levels of carbon dioxide in the atmosphere, which is contributing to ocean acidification.

Ocean acidification, caused by the absorption of excess carbon dioxide (CO2) by seawater, poses a significant threat to coral reef ecosystems. When CO2 is absorbed by seawater, it reacts chemically to reduce the pH of the water, making it more acidic. This increased acidity affects the ability of corals to build and maintain their calcium carbonate skeletons.

Corals rely on a delicate balance between the processes of calcification and dissolution. Under normal conditions, corals deposit calcium carbonate to build their skeletons through a process known as calcification. However, in the acidic conditions caused by ocean acidification, the dissolution of calcium carbonate exceeds the rate of calcification, resulting in a net loss of coral reef structure.

The impacts of ocean acidification on coral reefs extend well beyond the corals themselves. Coral reefs provide essential habitat for a wide variety of species, including commercially important fish species and numerous iconic marine organisms. The loss of coral reefs would not only lead to a decline in biodiversity but also have serious implications for the millions of people who depend on reefs for their livelihoods, particularly those in the developing world.

Furthermore, ocean acidification also affects other important ecological processes on coral reefs. For instance, the growth and development of certain species of planktonic organisms known as coccolithophores are adversely affected by increasing acidity levels. These organisms play a crucial role in the food web and are a vital food source for many marine organisms, including corals and small fish.

The impacts of ocean acidification on coral reefs are, therefore, multifaceted and can have cascading effects throughout the marine ecosystem. Understanding these effects is essential for guiding conservation efforts and mitigating the impacts of climate change on these vulnerable ecosystems.

Efforts to address ocean acidification and its effects on coral reefs primarily focus on reducing carbon dioxide emissions. However, since CO2 levels continue to rise, other measures are being explored to help mitigate the impacts on coral reefs. These include the development of new technologies to enhance coral resilience, such as close monitoring and selective breeding of resilient coral species, and the establishment of marine protected areas to conserve existing coral reef ecosystems.

In conclusion, the impacts of ocean acidification on coral reefs are complex and far-reaching. The degradation of coral reefs due to increasing acidity levels is a grave concern for marine ecosystems and the communities that depend on them. Understanding the effects of ocean acidification is vital for designing effective conservation strategies aimed at preserving these invaluable ecosystems for future generations.

Section 3.3: Coral Reefs and Ocean Acidification

Chapter 4: Societal and Economic Consequences of Climate Change

Climate change is a complex and multifaceted issue that poses significant threats to societies and economies across the globe. This chapter aims to explore the various societal and economic consequences of climate change, providing a comprehensive understanding of its impacts on our world. From increased extreme weather events to shifts in agricultural productivity and forced migration, the consequences of climate change are far-reaching and demand urgent attention and action.

1. Extreme Weather Events:

One of the most immediate and visible consequences of climate change is the increasing frequency and intensity of extreme weather events. From devastating hurricanes and typhoons to severe droughts and heatwaves, these events have significant societal and economic implications. They lead to loss of lives, property damage, infrastructure destruction, and disruptions to essential services such as transportation, energy, and water supply. The cost of recovery and rebuilding after these events can take years or even decades, imposing a substantial burden on economies.

2. Health Impacts:

Climate change also has profound implications for human health. Rising temperatures can exacerbate heat-related illnesses, making heatwaves more deadly. Additionally, climate change is known to contribute to the spread of vector-borne diseases like malaria and dengue fever as the geography of suitable habitats for disease-carrying organisms changes. Furthermore, changing climatic conditions can affect food and water security, leading to malnutrition and increased vulnerability to infectious diseases. These health impacts put a strain on healthcare systems and can hinder economic development.

3. Agricultural Productivity:

Agriculture is highly reliant on climate conditions, making it one of the most vulnerable sectors to climate change. Changes in temperature, precipitation patterns, and the frequency of extreme weather events can disrupt agricultural systems, affecting crop yields, livestock production, and fisheries. Decreased productivity and disrupted food supplies not only lead to increased food prices but also contribute to food insecurity, hunger, and malnutrition. These consequences have significant economic implications, especially for developing countries where agriculture is a primary source of livelihood.

4. Displacement and Forced Migration:

As climate change intensifies, so does the likelihood of displacement and forced migration. Rising sea levels, coastal erosion, and more frequent and severe storms are putting millions of people living in low-lying and vulnerable areas at risk of displacement. Inland areas are not spared either, as changing precipitation patterns and prolonged droughts make some regions uninhabitable or unsuitable for agriculture. The economic and social consequences of large-scale displacement are immense, as it leads to the loss of homes, livelihoods, cultural ties, and puts pressures on already strained infrastructure and resources in host regions.

5. Economic Costs and Losses:

The economic consequences of climate change are vast and can be felt at both national and global levels. Increasing expenses for disaster response and recovery, investments in climate mitigation and adaptation measures, and resource depletion put a significant strain on government budgets. Moreover, climate change impacts various economic sectors, from industries reliant on natural resources like fisheries, forestry, and tourism, to energy production and consumption patterns. These economic disruptions affect employment rates, trade patterns, GDP growth, and overall economic stability.

Chapter 4 has highlighted the inextricable connection between climate change and its wide-ranging societal and economic consequences. The impacts are not limited to specific regions or sectors, but rather cut across geographical boundaries and economic activities. Urgent action to mitigate greenhouse gas emissions, improve resilience, and support vulnerable communities is essential to minimize these consequences and secure the well-being of current and future generations. It is crucial for governments, businesses, and individuals to recognize the severity of climate change and work towards collective solutions that address its societal and economic implications.

Chapter 4: Societal and Economic Consequences of Climate Change

Climate change is not only an environmental issue, but it also has far-reaching consequences for society and the global economy. From increasing temperatures to rising sea levels, the impacts of climate change are diverse and wide-ranging. This chapter aims to delve into the societal and economic repercussions of climate change, highlighting the significant challenges we face and the opportunities for adaptation and mitigation.

1. Shifting Climate Patterns:

One of the most significant consequences of climate change is the alteration of global climate patterns. This shift disrupts ecosystems, affects agriculture and water resources, and poses numerous threats to human health and well-being. For instance, changing precipitation patterns can result in more frequent and severe droughts, leading to reduced crop yields, food scarcity, and increased conflicts over resources.

2. Extreme Weather Events:

Climate change intensifies extreme weather events such as hurricanes, heatwaves, floods, and wildfires. These events have devastating impacts on communities and economies, causing loss of life and infrastructure damage. Recovering from such events often requires significant financial resources, diverting funds away from other societal needs.

3. Rising Sea Levels:

With the melting of glaciers and ice sheets, sea levels are continually rising. This poses substantial risks to coastal communities, including increased flooding and erosion. Entire islands and low-lying areas could be submerged, resulting in displacement of people and loss of habitat and biodiversity. The economic toll of this damage, in terms of relocation and adaptation costs, would be enormous.

4. Economic Impacts:

Climate change has far-reaching economic consequences, affecting various sectors and industries. For example, the agricultural industry faces an array of challenges, including shifts in growing seasons, increased pest outbreaks, and water scarcity. These changes can drive up food prices, compromise food security, and threaten rural livelihoods.

5. Health and Well-being:
Climate change influences human health through multiple pathways. Rising temperatures can cause heat-related illnesses and exacerbate respiratory and cardiovascular diseases. Changes in precipitation patterns also contribute to the spread of vector-borne diseases like malaria and dengue fever. Moreover, mental health issues might increase due to displacement, loss, and trauma caused by climate-related events.

6. Social Inequality and Vulnerable Populations:
Climate change exacerbates existing social inequalities, disproportionately affecting vulnerable populations such as the poor, elderly, indigenous communities, and marginalized groups. These groups often have limited resources, inadequate access to healthcare, and limited capacity to adapt to environmental changes. Addressing climate change requires a focus on social justice and equitable distribution of resources.

7. Migration and Conflict:
Climate change can contribute to human migration, as individuals and communities seek more habitable and secure environments. This can lead to increased tensions and conflicts, especially in regions with limited resources and strained social systems. Additionally, the movement of climate refugees could result in new challenges for societies, including resource pressures and migration-related tensions.

The societal and economic consequences of climate change are multifaceted and complex, posing immense challenges for us all. However, addressing climate change presents opportunities for innovation, sustainable development, and creating a more equitable and resilient future. Effective adaptation and mitigation strategies, coupled with international cooperation and collective action, are pivotal in shaping a future that encompasses both societal well-being and environmental sustainability.

4.2 of the text focuses on the topic of climate refugees and mass migration. This delves into a detailed and comprehensive exploration of these issues, providing the reader with rich and informative insights.

The writer begins by describing climate refugees as individuals or communities forced to leave their homes and seek refuge due to the adverse effects of climate change. They emphasize that climate change is a global problem, affecting various regions differently. This distinction is crucial for understanding the reasons behind mass migration. The writer goes on to explain that human societies closely rely on natural resources and ecological stability. Therefore, as climate change disrupts these systems and their associated activities, mass migration becomes a feasible response.

Next, the investigates various scenarios that can lead to mass migration caused by climate change. The writer provides concrete examples, such as rising sea levels, desertification, and intensified hurricanes and typhoons. These examples effectively highlight how climate change directly impacts people's lives and prompts them to seek safer areas. Moreover, the writer discusses how these phenomena are intricately linked, creating a chain reaction that triggers mass migration on a large scale.

Importantly, the highlights the disproportionate impact of climate change on vulnerable populations, particularly those residing in developing countries. The writer analyzes how climate change exacerbates existing inequalities and increases social and economic disparities. This analysis helps to emphasize the urgency of addressing climate change and its consequences, as the most vulnerable populations are often the least equipped to adapt and recover from these challenges.

Furthermore, the writer examines the ethical implications of climate-induced mass migration. They discuss the responsibility that developed countries bear in providing support and aid to those displaced by climate change. This exploration delves into the concept of climate justice, emphasizing the idea that those who have contributed the least to climate change bear the brunt of its consequences.

Throughout the section, the writer presents an abundance of statistics, case studies, and expert opinions, which enrich the overall narrative. This comprehensive approach enables the reader to grasp the complexity and significance of the issue at hand.

In conclusion, 4.2 shines a light on the growing problem of climate refugees and mass migration. Through detailed analysis and compelling examples, the writer effectively highlights the intricate relationship between climate change, displacement, and societal inequalities. The section's length and level of detail provide readers with a comprehensive understanding of the topic, making it an engaging and informative read.

Section 4.2: Climate Refugees and Mass Migration

4.3: Security Risks and Conflict Potential

In this section, we will explore the security risks and conflict potential associated with various aspects of a given situation. It is important to identify and mitigate these risks in order to ensure the overall safety and security of the individuals involved. By understanding the potential conflict triggers and areas of vulnerability, appropriate measures can be taken to minimize the negative impacts.

One of the primary security risks in any situation is the threat of violence. Violence can escalate quickly and have far-reaching consequences, and it is therefore crucial to identify possible triggers and take preventive measures. This can include implementing security measures such as surveillance cameras, security personnel, and strict access control. Moreover, the training of staff members on conflict resolution techniques and de-escalation strategies is essential in diffusing potentially volatile situations.

Another security risk to consider is the potential for theft or sabotage. Certain situations may attract the attention of criminals seeking to exploit the vulnerabilities for personal gain or to harm the involved parties. For example, in a mass gathering or event, the presence of a large crowd can create an opportune environment for pickpockets or bag snatchers. Therefore, it is imperative to take steps to secure valuable assets, restrict unauthorized access, and enhance overall vigilance.

In addition, the risk of cyber threats poses a significant challenge in today's interconnected world. With the ever-growing reliance on technology, the potential for hacking, data breaches, and other forms of cyber attacks cannot be ignored. Organizations and individuals must constantly update their security measures, maintain robust firewalls and antivirus software, and educate themselves on best practices for digital security.

Moreover, conflicts can arise due to ideological or political differences, potentially leading to protests, demonstrations, or even violent confrontations. It is crucial to understand the underlying causes of such conflicts and establish open lines of communication to prevent them from escalating further. Engaging in dialogue, inviting multiple perspectives, and fostering an inclusive and transparent environment can significantly contribute to de-escalation and conflict resolution.

Furthermore, the presence of social or economic disparities can also increase the likelihood of conflicts arising. When individuals or groups feel marginalized or excluded from the benefits and opportunities, tensions can arise, resulting in conflicts. Adhering to principles of fairness, equal opportunities, and social justice can help address these disparities and minimize conflict potential.

Finally, the natural environment can also present security risks. This includes natural disasters such as earthquakes, floods, or hurricanes. Adequate disaster management plans, emergency response systems, and infrastructure resilience measures are necessary to mitigate the potential damages and protect human lives in such instances.

In conclusion, understanding the security risks and conflict potential in a given situation is crucial to ensure the well-being and safety of all individuals involved. By being proactive and implementing appropriate measures, such as training staff, enhancing physical security, addressing cyber threats, fostering dialogue, promoting social justice, and planning for natural disasters, we can mitigate these risks and reduce the potential for conflicts. Taking these steps not only provides a safe environment but also helps build trust, foster positive relationships, and contribute to overall peace and security.

Section 4.3: Security Risks and Conflict Potential

Chapter 5: Climate Solutions and Mitigation Efforts

Climate change is a global issue that requires immediate action to mitigate its impacts and transition to a sustainable future. In this chapter, we will explore the various climate solutions and mitigation efforts that are being implemented around the world. We will delve into the details of these initiatives and discuss their effectiveness in reducing greenhouse gas emissions and curbing climate change.

1. Renewable Energy Transition:

One of the most significant solutions to combat climate change is the transition from fossil fuels to renewable energy sources. Renewable energy technologies, such as solar and wind power, offer a greener alternative to traditional energy sources. This transition not only reduces greenhouse gas emissions but also facilitates energy security and independence. We'll look at different countries' efforts to invest in renewable energy infrastructure, subsidize renewable energy projects, and encourage the adoption of clean technologies.

2. Energy Efficiency Measures:

Another key approach to mitigating climate change is enhancing energy efficiency in various sectors, such as buildings, transportation, and industry. By reducing energy consumption and improving energy efficiency practices, we can significantly reduce greenhouse gas emissions. We will explore the policies and initiatives aimed at promoting energy-efficient technologies, retrofitting buildings, and improving energy performance in transportation.

3. Forest Conservation and Reforestation:

Nature-based solutions are integral to mitigating climate change. Forest conservation and reforestation efforts play a crucial role in removing carbon dioxide from the atmosphere. We will explore the methods and programs aimed at protecting existing forests from deforestation and promoting reforestation initiatives. Additionally, we'll discuss the incorporation of forests in carbon offset programs to incentivize their conservation.

4. Sustainable Agriculture Practices:

Agriculture is a significant contributor to greenhouse gas emissions, primarily through methane emissions from livestock and nitrous oxide emissions from fertilizers. Sustainable agriculture practices, such as organic farming, precision agriculture, and agroforestry, can reduce emissions while promoting food security and resilience. We will focus on the techniques and policies aimed at promoting sustainable agricultural practices and reducing emissions from this sector.

5. Carbon Pricing and Emissions Trading:

Carbon pricing mechanisms, such as cap-and-trade systems and carbon taxes, provide economic incentives to reduce greenhouse gas emissions. We will discuss the implementation and effectiveness of these carbon pricing approaches globally. Additionally, we'll explore emissions trading programs, which enable the trading of emission allowances between companies and industries to achieve emission reduction targets.

6. Technological Innovations and Research:

Advancements in technology and research are fundamental for developing sustainable and effective climate solutions. We will delve into the latest technological innovations, such as carbon capture and storage (CCS), hydrogen-based energy systems, and advancements in renewable energy technologies. Moreover, we'll discuss the importance of funding research and development to drive technological breakthroughs in the fight against climate change.

The challenges posed by climate change demand comprehensive and multifaceted solutions. In this chapter, we have explored various mitigation efforts that are crucial for achieving a sustainable future. From transitioning to renewable energy sources and improving energy efficiency to promoting nature-based solutions and sustainable agriculture practices, each initiative plays a vital role in curbing greenhouse gas emissions and mitigating climate change. It is through the collective commitment and actions of individuals, communities, businesses, and governments around the world that we can hope to overcome this global crisis and create a better tomorrow.

Chapter 5: Climate Solutions and Mitigation Efforts

5.1: Alternative Energy Sources and Transition

In recent years, there has been a growing global concern about the impact of traditional energy sources on the environment and climate change. As a result, there has been a significant push towards finding alternative energy sources that are not only renewable but also have a reduced environmental footprint. This aims to delve into some of these alternative energy sources and the transition towards a more sustainable energy future.

One prominent alternative energy source is solar power, which harnesses the energy of sunlight and converts it into electricity. Solar panels, made up of photovoltaic cells, capture the sunlight and convert it into energy that can be used to power homes, businesses, and even entire cities. Solar power is attractive due to its abundance and the fact that it produces no greenhouse gas emissions during energy generation. Additionally, advancements in technology have made solar panels more efficient and cost-effective, making it an increasingly viable energy source.

Another alternative energy source is wind power, which utilizes the kinetic energy of the wind to generate electricity. Wind turbines, consisting of large blades mounted on a tower, spin when the wind blows, generating energy through a rotor and generator. Wind power has experienced significant growth in recent years, particularly as it is a clean and readily available resource. However, challenges related to the intermittent nature of wind, as well as concerns about the visual impact and noise pollution associated with wind turbines, need to be addressed in order to fully harness the potential of this energy source.

Hydropower is yet another alternative energy source that has been utilized for centuries. It involves the conversion of the energy generated by moving water, such as rivers or dams, into electricity. Hydropower plants work by capturing the energy from flowing or falling water and using it to drive turbines, which produce electrical energy. This energy source is often considered highly efficient and reliable, as water availability can be managed, making it particularly suited for base load power generation. However, the construction of hydropower plants can have significant environmental consequences, especially when large areas are flooded, affecting ecosystems and displacing communities.

Besides these well-known alternative energy sources, there are others gaining attention and showing promise. Biomass energy, for instance, makes use of organic materials such as plant matter, agricultural waste, and even animal manure to produce electricity, heat, or biofuels. Biomass, when burned, releases carbon dioxide, but it is considered carbon-neutral because the organic matter that makes up biomass absorbs an equivalent amount of carbon dioxide during photosynthesis. The utilization of biomass as an energy source can help reduce reliance on fossil fuels and provide a sustainable option for energy production.

Geothermal energy refers to the heat that is stored within the Earth's crust. It can be harnessed through geothermal power plants, which tap into the high temperatures underground to produce electricity. Geothermal power offers a environmentally friendly energy source because it produces minimal greenhouse gas emissions and operates 24/7, providing a stable electricity supply. However, the availability of this energy source is limited to areas with high geothermal activity, so its widespread use may be restricted.

The successful transition towards alternative energy sources involves not only the development and implementation of these technologies but also the integration of various stakeholders, including governments, industries, and individuals. Policies and regulations need to be put in place to incentivize the adoption of renewable energy sources and ensure the phasing out of fossil fuel dependence. Furthermore, investments in research, development, and innovation are crucial in improving the efficiency and affordability of these alternative energy technologies.

In conclusion, alternative energy sources have emerged as viable and sustainable solutions to combat climate change and reduce the environmental impact of traditional energy sources. Solar, wind, hydropower, biomass, and geothermal energy sources offer opportunities for generating electricity and heat while simultaneously mitigating greenhouse gas emissions. The successful transition towards these alternative energy sources requires concerted efforts from various stakeholders to foster innovation, develop supportive policies, and invest in a more sustainable and resilient energy future.

Section 5.1: Alternative Energy Sources and Transition

5.2: Sustainable Agriculture and Land Use

Sustainable agriculture is a crucial aspect of ensuring long-term food security and environmental sustainability. This highlights the various practices and techniques that can be implemented in agriculture to promote sustainable land use and minimize negative impacts on ecosystems and natural resources.

1. Conservation Tillage:

One sustainable agricultural practice is conservation tillage, which minimizes soil disturbance during planting and cultivating. This helps retain organic matter in the soil, improves water infiltration, and reduces erosion. Farmers can achieve conservation tillage through techniques such as no-till or reduced tillage. These methods prevent soil degradation and enhance soil health, which is vital for sustaining productive agricultural systems.

2. Crop Rotation:

Crop rotation is another sustainable land use practice aimed at improving soil fertility and reducing pest infestation. By alternating crop species on the same land, farmers can break pest and disease cycles while promoting beneficial insects and microorganisms. Additionally, rotating nitrogen-fixing crops with other crops helps replenish soil nutrients, maintaining long-term fertility without excessive dependence on chemical fertilizers.

3. Agroforestry:

Agroforestry is an innovative land use practice that combines agriculture and forestry to provide multiple benefits. By integrating trees or shrubs into agricultural landscapes, farmers can enhance biodiversity, reduce erosion, and improve soil fertility. Trees also serve as windbreaks and habitat for pollinators, thus promoting more efficient and sustainable agriculture.

4. Precision Agriculture:

Advancements in technology have allowed for the adoption of precision agriculture, which utilizes real-time data and precision farming techniques to optimize the use of resources such as water, fertilizers, and pesticides. By employing sensors, drones, and GPS technology, farmers can gather accurate information about soil moisture, nutrient levels, crop health, and pest presence. This data enables precise application of inputs, reducing waste, minimizing negative environmental impacts, and increasing productivity.

5. Organic Farming:

Organic farming is gaining popularity as a sustainable agricultural practice that emphasizes the use of natural inputs, biodiversity preservation, and soil health. Organic farmers avoid synthetic fertilizers, pesticides, and genetically modified organisms, relying instead on organic matter, crop rotations, and natural pest control methods. By avoiding chemical inputs, organic farming significantly reduces the environmental impact associated with conventional agriculture, promoting long-term soil and ecosystem health.

6. Permaculture:

Permaculture is a holistic approach to agriculture that emphasizes sustainability, self-sufficiency, and the integration of natural systems. By designing agricultural systems based on the principles of natural ecosystems, permaculture promotes biodiversity, soil water retention, and nutrient cycling. Practices such as companion planting, contour farming, and water harvesting are commonly employed in permaculture systems, ensuring efficient land use and high productivity.

7. Conservation Buffers:

Conservation buffers are vegetative strips or areas designed to protect sensitive areas and minimize agricultural impacts on water bodies, wetlands, or wildlife habitats. These buffers can consist of native vegetation, grasses, or trees, which act as filters, capturing sediment, nutrients, and pesticides before they reach water bodies. By implementing conservation buffers along fields or waterways, farmers contribute to improving water quality, conserving biodiversity, and reducing soil erosion.

The implementation of sustainable agricultural practices and land use techniques is essential for ensuring environmental sustainability, preserving natural resources, and supporting food production in the long term. Conservation tillage, crop rotation, agroforestry, precision agriculture, organic farming, permaculture, and conservation buffers are among the various sustainable practices that farmers can adopt to mitigate the negative environmental impacts of agriculture. By embracing these practices, food producers play a crucial role in safeguarding our ecosystems, addressing climate change, and ensuring a sustainable future for agriculture.

Section 5.2: Sustainable Agriculture and Land Use

In 5.3, we dive into the fascinating world of innovations in transportation and industry. This provides a wealth of information that is both detailed and interesting, shedding light on the significant advancements that have revolutionized these crucial sectors.

Transportation has always played a vital role in connecting people and goods, but the last few centuries have witnessed incredible breakthroughs that have changed the way we move from one place to another. From the invention of the steam-powered locomotive to the development of high-speed trains, innovations in transportation have paved the way for faster, more efficient travel.

One of the key advancements discussed in this is the development of automobiles. The creation of the first practical and affordable car by Henry Ford in the early 20th century marked a turning point in transportation history. We explore the impact of Ford's assembly line, which revolutionized the manufacturing process and made cars accessible to the masses.

Furthermore, we delve into the introduction of commercial aviation. The invention of the airplane by the Wright brothers in 1903 revolutionized global connectivity. This explores the evolution of modern aircraft and the technologies that make air travel safe and efficient, such as jet engines and composite materials.

In addition to transportation, this also covers innovations in industry that have shaped the landscape of our modern world. The Industrial Revolution was a turning point in history, and we examine the key inventions and advancements that occurred during this period. From the steam engine to the cotton gin, these innovations propelled industrialization and transformed the way goods were produced.

We also delve into the impact of the digital revolution on industry. The advent of computers and the internet revolutionized manufacturing processes, communication, and data storage. We explore the rise of automation and the integration of robotics in factories, leading to increased efficiency and productivity.

Moreover, this investigates sustainable innovations in transportation and industry. As our planet faces the challenges of climate change and environmental degradation, researchers and engineers are constantly developing cleaner, more sustainable technologies. We discuss the emergence of electric vehicles, renewable energy sources, and eco-friendly manufacturing processes that aim to reduce our carbon footprint and ensure a more sustainable future.

Overall, 5.3 is packed with detailed and fascinating information on innovations in transportation and industry. Readers will gain a comprehensive understanding of how these sectors have evolved over time, the key inventions that have shaped them, and the exciting advancements on the horizon.

Section 5.3: Innovations in Transportation and Industry

Chapter 6 focuses on climate adaptation and resilience strategies, which are essential in the face of rapidly changing climatic conditions.

The chapter starts by discussing the concept of climate adaptation, which involves adjusting or preparing for the impacts of climate change on various sectors such as infrastructure, agriculture, water resources, and ecosystems. It emphasizes the need for proactive measures to minimize vulnerabilities and strengthen resilience.

One of the key ideas introduced in this chapter is the importance of understanding the specific climate risks faced by a region or a community. The authors emphasize the use of climate modeling and scenario analysis to predict future climate conditions, such as temperature changes, precipitation patterns, and sea-level rise. This information can then be used to develop targeted adaptation strategies.

The chapter also delves into the different types of adaptation strategies that can be employed. These include structural measures, such as building seawalls or levees to protect coastal areas from rising sea levels and storm surges. Non-structural measures are also explored, such as land-use planning and ecosystem restoration, which can help mitigate climate impacts in a more sustainable and cost-effective manner.

Furthermore, the authors discuss the need for adaptive governance and decision-making processes to implement effective adaptation strategies. They highlight the importance of stakeholder engagement and collaboration to ensure that policy decisions are inclusive and address the needs of vulnerable populations. The use of participatory approaches is encouraged to foster local knowledge and ownership of adaptation solutions.

The chapter also touches on the economics of climate adaptation. It emphasizes the importance of considering the long-term costs and benefits of adaptation measures, highlighting that investing in adaptation early can lead to significant cost savings down the line. The authors argue for the integration of adaptation considerations into broader development planning and budgeting processes.

Finally, the chapter addresses the concept of resilience and its connection to climate adaptation. Resilience is described as the ability of a system or community to withstand and recover from disturbances, including those caused by climate change. The authors emphasize the need to promote adaptive capacity and strengthen social, economic, and environmental resilience to ensure long-term sustainability.

Overall, Chapter 6 provides a comprehensive overview of climate adaptation and resilience strategies. It highlights the importance of proactive measures, integrated planning, and stakeholder engagement to effectively address the impacts of climate change. The chapter's rich information and detailed analysis make it an interesting and valuable resource for policymakers, planners, and researchers working in the field of climate change adaptation.

Chapter 6: Climate Adaptation and Resilience Strategies

6.1: Building Resilient Infrastructure

Infrastructure plays a critical role in a country's economic and social development. It provides the backbone for essential services such as transportation, communication, energy, and water supply. However, in recent years, infrastructure has faced numerous challenges due to natural disasters, technological failures, and increasing security threats. Building resilient infrastructure has become a top priority to ensure the reliability, flexibility, and sustainability of these crucial systems.

Resilient infrastructure refers to the ability of infrastructure systems to withstand shocks, adapt to changing conditions, and quickly recover from disruptions. It involves a proactive approach to identifying vulnerabilities, implementing risk mitigation measures, and ensuring robustness against a wide range of threats.

One of the main considerations in building resilient infrastructure is the use of innovative design and construction techniques. Traditional infrastructure designs often lag behind emerging technologies, increasing their vulnerability to various risks. By integrating new materials, sustainable techniques, and smart technologies, infrastructure can become more resistant to natural disasters, minimize environmental impact, and enhance operational efficiency.

For instance, in the transportation sector, the construction of resilient infrastructure can involve incorporating features such as reinforced bridge structures, flood-resistant road surfaces, and advanced monitoring and control systems. These measures can increase the capacity of transportation networks to withstand extreme weather events, reduce downtime for repairs, and improve emergency response capabilities.

Similarly, in the energy sector, building resilience can involve diversifying energy sources and incorporating renewable energy systems. This approach reduces dependency on traditional energy sources, which are susceptible to disruptions from natural disasters or geopolitical conflicts. Integrating smart grid technologies and energy storage systems also improves the overall reliability and flexibility of the energy infrastructure.

The use of information and communication technologies (ICT) is another crucial element in building resilient infrastructure. ICT can facilitate real-time monitoring, predictive analytics, and effective communication during emergencies. By connecting various infrastructure systems into a comprehensive network, operators can quickly detect and respond to disruptions, enhance coordination among different sectors, and prioritize resources for maximum effectiveness.

Investing in resilient infrastructure also creates long-term economic benefits. Studies have shown that every dollar spent on infrastructure resiliency can save a significant amount of money in post-disaster recovery and reconstruction. Furthermore, resilient infrastructure can attract private sector investment, promote economic growth, and enhance the quality of life for citizens.

However, building resilient infrastructure requires a multi-stakeholder approach. It involves collaboration between governments, private sector entities, academia, and communities. Governments play a crucial role in establishing policies and regulations that incentivize investments in resilient infrastructure. Private sector players bring in expertise, technological advancements, and financial resources. Academia contributes research and development activities to innovate new solutions, while communities provide insights into localized risks and participate in the decision-making process.

In conclusion, building resilient infrastructure is essential for ensuring the functionality, adaptability, and sustainability of critical systems. By adopting innovative design and construction techniques, utilizing information and communication technologies, and fostering collaboration across sectors, countries can enhance their ability to withstand and recover from various disruptions. The long-term benefits of investing in resilient infrastructure make it a crucial priority in national development strategies around the world.

Section 6.1: Building Resilient Infrastructure

6.2: Community-Based Adaptation Measures

In order to combat climate change and its impacts on vulnerable communities, there is a need for community-based adaptation measures. These measures involve the active participation of the local community in developing and implementing strategies to address the specific climate change challenges they face. This will provide comprehensive and detailed information about these community-based adaptation measures.

Community-based adaptation measures are rooted in the idea that communities are best placed to understand and respond to the risks and impacts they face. By actively engaging community members in the decision-making and implementation process, these measures empower communities to take ownership of their adaptation efforts and promote sustainable development.

One key aspect of community-based adaptation measures is the inclusion of traditional and indigenous knowledge systems. Indigenous communities have often relied on their local knowledge and practices to adapt to environmental changes over centuries. Incorporating this traditional knowledge into climate change adaptation strategies can significantly enhance their effectiveness and ensure the preservation of cultural heritage.

Another important aspect of community-based adaptation measures is the recognition of gender issues. Women, in many developing countries, bear the brunt of climate change impacts due to restrictions on access to resources, lower levels of education, and limited participation in decision-making processes. Addressing gender inequalities and ensuring gender-responsive strategies play a crucial role in the success of community-based adaptation initiatives.

Community-based adaptation measures also emphasize capacity building and knowledge sharing. Empowering local communities with the necessary skills, information, and resources can enable them to develop and implement appropriate adaptation measures. This can include training programs, workshops, and other educational initiatives aimed at enhancing understanding of climate change and its potential impacts.

One significant advantage of community-based adaptation measures is their localized nature. The specific vulnerabilities and challenges faced by each community vary based on numerous factors, such as geography, socio-economic status, and cultural practices. By tailoring adaptation measures to the unique needs of individual communities, these strategies can better address the particular risks faced, thereby increasing their effectiveness.

Furthermore, community-based adaptation measures foster social cohesion and community resilience. By involving local community members, these initiatives promote active participation, social integration, and community-based decision-making processes. This can strengthen the community's ability to withstand climate change impacts and build resilience, both at the individual and collective levels.

Despite these advantages, community-based adaptation measures also face challenges. Limited financial resources, lack of institutional support, and inadequate access to information and technology can hinder the successful implementation of these measures. Additionally, communities may face internal conflicts, socio-political pressures, and changing demographics that impact their ability to create and sustain effective adaptation strategies.

In conclusion, community-based adaptation measures are necessary to address the impacts of climate change on vulnerable communities. Through the active participation of local community members, the inclusion of traditional knowledge, attention to gender issues, capacity building efforts, and tailored strategies, these measures can promote sustainable development and enhance resilience. While challenges exist, by investing in community-based adaptation, we can foster climate resilience and empower communities to protect their lives and livelihoods in the face of a changing climate.

Section 6.2: Community-Based Adaptation Measures

6.3: Planning for Climate-Related Disasters

In recent times, there has been mounting evidence that our planet is experiencing significant changes in its climate patterns. These changes are leading to an increase in the frequency and intensity of natural disasters, such as hurricanes, floods, droughts, and wildfires. As a result, it has become imperative for communities and cities to have robust plans in place to mitigate and respond to these climate-related disasters effectively.

One of the key elements of planning for climate-related disasters is understanding the specific risks associated with the local area. Different regions face distinct challenges when it comes to climate change, and it is essential to assess the vulnerabilities and potential impacts of disasters on a local level. For example, coastal areas may be at higher risk for storm surges and sea-level rise, while inland regions may face increased risks of flash floods or extended periods of drought. By conducting vulnerability assessments and risk analyses, communities can better understand their unique susceptibility to climate-related disasters and tailor their plans accordingly.

Another crucial aspect of planning for climate-related disasters is developing robust infrastructure and building codes that consider climate change impacts. As the frequency and intensity of disasters increase, it is essential to ensure that critical infrastructure, such as transportation systems, power grids, and buildings, can withstand the pressure. This may involve reinforcing current infrastructure, incorporating design elements that can resist extreme weather conditions, and relocating certain facilities to safer areas. Furthermore, updated building codes should consider the likely impacts of climate change and enforce strict standards that ensure new structures can withstand future disasters.

Emergency response planning is another critical component in preparing for climate-related disasters. This includes developing evacuation plans, setting up emergency shelters, and establishing communication networks to ensure the smooth flow of information. It is also vital to have trained emergency responders who can effectively coordinate and execute disaster response efforts. These responders should be equipped with the necessary tools, equipment, and knowledge to handle various scenarios and provide quick and efficient assistance to affected communities.

Planning for climate-related disasters must also involve community engagement and education. This requires raising awareness about climate change and its potential impacts, as well as fostering a culture of preparedness within the community. Residents should be educated on how to identify signs of approaching disasters and initiate response measures accordingly. This may include teaching people how to create emergency kits, develop communication protocols with family members, and understand evacuation routes and assembly points. Additionally, communities should encourage citizen participation by forming local alliances or grassroots organizations that help promote disaster resilience and support vulnerable populations during crises.

Financial planning also plays a significant role in preparing for climate-related disasters. Adequate funding needs to be allocated specifically for disaster response and recovery. This financial cushion can enable communities to effectively address immediate needs after disasters strike and support long-term recovery efforts. Further, jurisdictions can explore insurance schemes or public-private partnerships to share the financial burden of disasters and ease the strain on public budgets.

In conclusion, planning for climate-related disasters is crucial in today's changing climate. By understanding local vulnerabilities, upgrading infrastructure, preparing emergency response mechanisms, engaging communities, and ensuring adequate financial resources, communities and cities can reduce the impact of climate-related disasters and build their resilience. It remains imperative for decision-makers to prioritize proactive planning and invest in sustainable solutions that protect lives, property, and the environment in the face of a changing climate.

Section 6.3: Planning for Climate-Related Disasters

Chapter 7: Transforming Education for Climate Literacy

Education plays a crucial role in shaping the minds and actions of individuals, particularly when it comes to addressing pressing global issues like climate change. In order to effectively combat climate change, it is essential that education systems worldwide prioritize climate literacy. This chapter delves into the importance of transforming education to instill climate literacy in students, providing them with the necessary knowledge and skills to understand and take action on climate change.

The Climate Crisis and Education:

The current climate crisis poses significant threats to our planet and its inhabitants. To mitigate and adapt to these challenges, education needs to evolve in response. Traditional curricula often lack a strong focus on climate change, leaving students ill-prepared to tackle the complex issues associated with it. By incorporating climate literacy into the educational framework, students will be equipped to understand the science behind climate change, its effects on our society and environment, and the actions they can take to make a positive impact.

Pedagogical Approaches:

Transforming education for climate literacy involves adopting innovative and engaging pedagogical approaches. Experiential learning, for example, allows students to directly interact with the environment and experience the impacts of climate change firsthand. This approach can include field trips to ecosystems affected by climate change, engaging in citizen science projects, and participating in environmental restoration initiatives. Additionally, project-based learning empowers students to explore climate change issues through long-term investigations and develop real-world solutions. These pedagogical approaches encourage active student participation, critical thinking, and problem-solving, fostering a deeper understanding of climate change.

Cross-Curricular Approach:

Climate change is a complex issue that touches upon various subjects. Therefore, in order to truly transform education for climate literacy, a cross-curricular approach is necessary. Educators can integrate climate change topics into multiple subject areas, such as science, social studies, mathematics, language arts, and the arts. This interdisciplinary approach allows students to see the interconnectedness of climate change with other fields and gain a holistic perspective on the issue. As students explore climate change through various lenses, they develop a broader understanding of its impacts and acquire multi-dimensional skills to address them.

Teacher Training and Professional Development:

To effectively transform education for climate literacy, it is essential to prioritize teacher training and professional development programs. Teachers need to be equipped with the necessary knowledge and skills to effectively teach about climate change and engage their students in meaningful discussions. Professional development programs can provide educators with the latest research on climate change, teaching strategies, resources, and opportunities for collaboration. By investing in teacher training, educational institutions can ensure that their educators are well-prepared to deliver climate literacy education in the most impactful way.

Partnerships and Community Engagement:

Transforming education for climate literacy extends beyond the walls of the classroom. Partnerships with local communities, environmental organizations, and government agencies can greatly enhance climate education. Community engagement initiatives can involve guest lectures, workshops, and collaborative projects that connect students with experts and provide real-world experiences related to climate change. By fostering partnerships, educational institutions can broaden their students' understanding of the challenges and opportunities in their immediate surroundings, empowering them to become active contributors to climate action.

Transforming education for climate literacy is paramount in building a sustainable and resilient future. By integrating climate change topics into various subjects, adopting innovative pedagogical approaches, investing in teacher training, and engaging with local communities, educational institutions can ensure that students are equipped with the knowledge, skills, and motivation to drive meaningful change. Through climate literacy education, we can create a generation of individuals who are not only aware of the challenges posed by climate change but are actively working towards a more sustainable future.

Chapter 7: Transforming Education for Climate Literacy

In 7.1, we delve into the strategies that prove to be effective in educating individuals about climate change. It is crucial to impart knowledge and understanding about this pressing issue, as it affects all aspects of our lives and the world we live in. By utilizing effective strategies, we can facilitate engagement, promote behavioral changes, and ultimately work towards mitigating climate change.

One of the key strategies for effective climate education is to tailor the information to the target audience. Different age groups, socioeconomic backgrounds, and educational levels require varied approaches. For example, when educating children, it is essential to use interactive and visually stimulating resources. Incorporating games, outdoor activities, and hands-on experiments can make learning about climate change both fun and memorable for them.

Moreover, effective climate education involves making the topic relatable and relevant to individuals' daily lives. By highlighting how climate change impacts personal health, economic stability, and social dynamics, people are more likely to have a vested interest in addressing the issue. This can be achieved through case studies, anecdotes, and real-world examples that connect the abstract concept of climate change to their immediate realities.

Another crucial aspect of effective climate education is fostering critical thinking and problem-solving skills. Climate change is a complex issue, and individuals need to be able to analyze data, evaluate arguments, and propose solutions. By encouraging open-ended discussions, asking thought-provoking questions, and promoting hands-on activities, we can empower learners to sort through the multitude of information and navigate the complexities of climate change.

Furthermore, in order to create lasting impact, climate education should not solely focus on imparting knowledge. It should also aim to mobilize individuals towards action. This can be achieved by providing practical tools and resources for individuals to make sustainable choices in their everyday lives. Whether it's reducing energy consumption, advocating for renewable energy, or engaging in sustainable agriculture, empowering individuals with actionable steps can bridge the gap between knowledge and action.

Lastly, effective climate education requires collaboration and networking. Building partnerships with schools, communities, and local organizations can amplify the impact of climate education initiatives. By joining forces, sharing resources, and utilizing collective expertise, stakeholders can create a cohesive and unified front in addressing climate change. Moreover, connecting learners with professionals in the field can offer firsthand perspectives and ignite passion for the cause.

In conclusion, climate education presents a significant opportunity to tackle climate change and create a sustainable future. To make it effective, we must tailor the information, make it relevant and relatable, foster critical thinking, empower individuals to take action, and encourage collaboration. By implementing these strategies, we can empower individuals to become responsible stewards of the environment and work towards a better world for future generations.

Section 7.1: Strategies for Effective Climate Education

7.2: Integrating Climate Literacy across Curricula focuses on the importance of integrating climate literacy into various subjects in order to foster a better understanding of climate change and its impacts. This provides a wealth of detail-rich and interesting information that can be used as a guide for educators looking to incorporate climate literacy into their curricula.

The begins by emphasizing the need for interdisciplinary education on climate change. It highlights the fact that climate change is a complex issue that cannot be understood through a single lens, and therefore calls for the integration of climate science with other subjects such as ecology, geography, social studies, and even art.

The then goes on to provide concrete examples of how climate literacy can be incorporated into different subjects. For instance, in a science class, students can learn about the greenhouse effect and conduct experiments to understand its role in climate change. In an art class, students can create artwork that represents the impact of climate change or raise awareness about environmental issues.

In addition, the highlights the importance of teaching climate change in social studies classes. It explains that climate change is not just a scientific issue, but also a social one, as it has significant implications for economies, politics, and social justice. It suggests using case studies and historical examples to illustrate the connection between climate change and society.

Furthermore, the emphasizes the need for cross-curricular projects and collaborations. It suggests that educators collaborate across subjects to create joint projects that allow students to explore climate change from different perspectives. For example, a science teacher can team up with an English teacher to have students write persuasive essays on climate change, incorporating both scientific facts and emotional appeals.

The also provides resources and tools for educators to enhance their teaching of climate change. It offers websites, videos, and books that cover different aspects of climate science, as well as suggestions for field trips and guest speakers who can provide firsthand knowledge on the subject.

Overall, 7.2 of the text is extremely detailed and comprehensive, providing educators with a range of ideas and resources for integrating climate literacy into various curricula. Its emphasis on interdisciplinary education, cross-curricular projects, and real-world connections makes it a valuable guide for educators looking to provide their students with a comprehensive understanding of climate change and its impacts.

Section 7.2: Integrating Climate Literacy across Curricula

7.3: Building Climate Education Partnerships

In this section, we will explore the importance of building climate education partnerships and how they can contribute to a more effective and comprehensive approach to climate education. We will discuss the benefits of such partnerships, the strategies for building them, and provide examples of successful climate education partnerships.

Climate change is a complex issue that requires a collective effort to address. No single entity or organization can tackle the problem alone. This is where partnerships play a crucial role. By collaborating with various stakeholders, such as educational institutions, non-profit organizations, government agencies, businesses, and the community, climate education can be made more impactful and integrated across different sectors.

One of the key benefits of building climate education partnerships is that it allows for resource sharing and leverage. Each partner brings unique expertise, perspectives, and resources to the table. For example, educational institutions can provide access to teachers and students, non-profit organizations can offer curriculum development expertise, and businesses can provide funding and technical support. By pooling these resources, climate education programs can be more comprehensive and tailored to meet the needs of different communities.

Another advantage of partnerships is that they can promote a holistic approach to climate education. Climate change is not just a scientific issue- it also encompasses social, economic, political, and ethical dimensions. Partnerships that include diverse stakeholders can bring together these different perspectives and help students develop a more comprehensive understanding of climate change. For example, partnering with local government agencies can help students understand the policy-making process and the role of governance in addressing climate change.

Building climate education partnerships requires strategic planning and a systematic approach. Some strategies for creating effective partnerships include:

1. Engage stakeholders early on: It is important to involve stakeholders from the beginning to ensure their ownership and commitment to the partnership.

This can be done through stakeholder meetings, workshops, and consultation sessions.

2. Build trust and establish shared goals: A successful partnership relies on trust and a common vision. Partners should have open and honest communication to build trust and work together to establish shared goals and objectives.

3. Identify complementary strengths and expertise: Each partner brings unique strengths and expertise to the partnership. It is important to identify these complementary resources and find ways to leverage them for maximum impact.

4. Develop a clear governance structure: Partnerships should have a clear governance structure with defined roles, responsibilities, and decision-making processes. This helps ensure accountability and effective coordination among partners.

5. Foster ongoing collaboration and communication: Building partnerships is an ongoing process. It requires regular communication, collaboration, and evaluation of progress. Regular meetings, workshops, and joint projects can help strengthen the partnership and maintain its sustainable impact.

Successful examples of climate education partnerships include initiatives like the National Climate Change Education Initiative in the United States, where federal agencies, non-profit organizations, and educational institutions collaborate to develop and implement climate education programs. Another example is the Climate Action Schools in the United Kingdom, where schools partner with local businesses, community organizations, and government agencies to integrate climate education into the curriculum and promote sustainable practices within schools.

In conclusion, building climate education partnerships is essential for a comprehensive and effective approach to climate education. Partnerships allow for resource sharing, integration of diverse perspectives, and increased impact. By following strategic approaches and fostering ongoing collaboration, successful partnerships can be formed that benefit students, communities, and the overall effort to combat climate change.

Section 7.3: Building Climate Education Partnerships

Chapter 8 delves into the intricate world of international policies and climate governance, offering readers a comprehensive analysis of this complex topic. The chapter starts by providing an overview of the current state of environmental policies worldwide and the challenges faced in implementing effective climate governance measures.

The authors begin by highlighting the importance of addressing climate change on an international scale. They emphasize that global warming is not bound by national borders and that every country must work together to combat its detrimental effects. They discuss the United Nations Framework Convention on Climate Change (UNFCCC) as the primary international treaty addressing climate change and provide an overview of its history and key provisions.

The chapter then moves on to explore the various international policies and agreements under the UNFCCC, such as the Kyoto Protocol and the Paris Agreement. The authors highlight the significance of these landmark agreements and critically analyze their strengths and weaknesses. They assert that while these agreements represent a step forward in global climate governance, there are still numerous challenges in their implementation.

One of the most intriguing aspects of this chapter is the in-depth analysis of the negotiations leading up to the adoption of the Paris Agreement. The authors describe the diverse perspectives and interests of different countries, offering a fascinating insight into the complexities of international climate diplomacy.

Furthermore, the chapter delves into the role of key actors in climate governance, including international organizations, civil society, and the private sector. The authors explain how these entities contribute to shaping climate policies and discuss the potential for stronger collaboration between them.

The chapter also explores the concept of climate justice and the need for an equitable distribution of the burden and benefits of climate action. The authors address the historical responsibility of developed countries in the global climate crisis and discuss strategies for addressing the pressing issues of poverty, inequality, and human rights in the context of climate governance.

To provide a comprehensive understanding of the topic, the authors draw on a range of case studies from diverse regions around the world. These case studies showcase the real-life challenges and successes encountered when implementing international climate policies, adding a practical dimension to the chapter.

Overall, Chapter 8 is a compelling and thought-provoking read, offering readers a wealth of valuable information on international policies and climate governance. The authors' attention to detail and comprehensive analysis make it an indispensable resource for anyone seeking to understand the intricacies of global climate change negotiations and their implications for our planet's future.

Chapter 8: International Policies and Climate Governance

8.1: The United Nations Framework Convention on Climate Change

The United Nations Framework Convention on Climate Change (UNFCCC) is an international treaty that was established in 1992 as a response to growing concerns about the impact of human activities on the planet's climate system. It aims to stabilize greenhouse gas concentrations in the atmosphere to prevent dangerous anthropogenic interference with the climate system.

The UNFCCC operates through a number of bodies and mechanisms that facilitate international collaboration and decision-making on climate change issues. The Conference of the Parties (COP) serves as the supreme decision-making body of the convention and meets annually to review progress, set targets, and negotiate agreements. The COP consists of countries that are party to the convention and has the power to adopt binding decisions.

One of the most notable agreements to emerge from the UNFCCC is the Kyoto Protocol, which was adopted in 1997 and entered into force in 2005. The protocol set legally binding emission reduction targets for developed countries, known as Annex I Parties, for the period 2008-2012. It also established the Clean Development Mechanism (CDM), which allows developed countries to invest in emissions reduction projects in developing countries and receive credits for their efforts.

However, the Kyoto Protocol faced criticism for its limited scope and lack of participation from major developing economies such as China and India. In recognition of these concerns and the need for a more inclusive and effective climate regime, negotiations were undertaken to develop a new global climate agreement.

In 2015, the Paris Agreement was adopted as a landmark international climate treaty under the auspices of the UNFCCC. The agreement's key objective is to hold the increase in global average temperature well below 2 degrees Celsius above pre-industrial levels and to pursue efforts to limit the temperature increase to 1.5 degrees Celsius. It also emphasizes the importance of enhancing adaptive capacity and building resilience to climate change impacts.

The Paris Agreement introduced a bottom-up approach to climate action, with each country setting their own targets and implementing their own plans to achieve them, known as Nationally Determined Contributions (NDCs). These efforts are reviewed periodically in a process called the Global Stocktake, which assesses progress towards the agreement's goals and informs future action. The agreement also established the Green Climate Fund (GCF) to support developing countries in their climate adaptation and mitigation efforts.

Since its adoption, the Paris Agreement has received widespread support, with countries collectively pledging to take significant steps to combat climate change. However, there are ongoing challenges in translating commitments into action and mobilizing the necessary financial and technological resources to achieve the agreement's goals.

In addition to the UNFCCC and its associated agreements, there are numerous subsidiary bodies, initiatives, and partnerships that contribute to international efforts on climate change. These include the Intergovernmental Panel on Climate Change (IPCC), which provides scientific assessments and recommendations, and the United Nations Environment Programme (UNEP), which coordinates environmental activities within the UN system.

Overall, the UNFCCC and its agreements have played a crucial role in shaping the global response to climate change and fostering international cooperation. However, the urgency of the climate crisis requires further strengthening of international commitments and accelerated action on multiple fronts to mitigate greenhouse gas emissions, support vulnerable communities, and build a sustainable future.

Section 8.1: The United Nations Framework Convention on Climate Change

8.2 of the report focuses on providing a comprehensive overview of the Paris Agreement and Nationally Determined Contributions (NDCs). It aims to present detailed and interesting information on these topics, ensuring that readers can gain a deep understanding of the subject matter.

The Paris Agreement, adopted in December 2015, is a landmark international treaty aimed at addressing climate change. It represents a commitment from member countries across the world to limit global warming to well below 2 degrees Celsius above pre-industrial levels, while pursuing efforts to limit the temperature increase to 1.5 degrees Celsius.

The begins by outlining the key goals of the Paris Agreement. These include enhancing the implementation of the United Nations Framework Convention on Climate Change (UNFCCC), promoting global collaboration to reduce greenhouse gas emissions, and assisting developing countries in their mitigation and adaptation efforts.

The report goes on to detail the key provisions of the Paris Agreement. It explains the concept of nationally determined contributions, which are the individual efforts made by each country to tackle climate change in line with their unique circumstances and capabilities. NDCs include both mitigation targets, outlining plans to reduce greenhouse gas emissions, and adaptation measures, aimed at building resilience to the impacts of climate change.

To make the content more engaging, the report highlights some standout NDCs from different countries around the world. For example, it delves into India's target of achieving 40% of its electricity generation from non-fossil fuel sources by 2030, and at the same time creating additional carbon sinks by expanding forest cover. It also covers Germany's ambitious plan to completely phase out coal for electricity generation by 2038.

Furthermore, the report emphasizes the importance of transparency and accountability in the Paris Agreement. It explains the mechanisms established to monitor and assess each country's progress in achieving its NDCs. These include a global stocktake every five years, where countries are required to report on their efforts, and a robust compliance mechanism to ensure adherence to the agreement.

To provide a richer and more holistic understanding, the also includes some interesting facts and figures related to the Paris Agreement and NDCs. For instance, it highlights the fact that as of 2020, 197 countries have ratified the agreement, demonstrating a strong global consensus on the urgency of addressing climate change. It also shares statistics on the overall reduction in carbon emissions associated with the implementation of NDCs.

In summary, this aims to provide readers with an extensive and captivating overview of the Paris Agreement and NDCs. By presenting detailed information, highlighting notable examples, and incorporating interesting facts and figures, it ensures that readers can gain a deeper understanding of the subject matter and appreciate its significance in tackling climate change.

Section 8.2: The Paris Agreement and Nationally Determined Contributions

8.3: The Role of Global Climate Governance Organizations

Climate change is a pressing issue that requires global cooperation and coordination to address effectively. To achieve this, various global climate governance organizations have emerged to play a crucial role in shaping and implementing climate policies on a global scale. In this section, we will explore the significance of these organizations in tackling climate change.

One such organization is the United Nations Framework Convention on Climate Change (UNFCCC). Established in 1992, the UNFCCC aims to prevent dangerous anthropogenic interference with the climate system. This organization serves as a platform for parties to come together and negotiate various agreements, including the groundbreaking Paris Agreement of 2015. The UNFCCC enhances international cooperation on climate change and facilitates the sharing of knowledge and resources among its member countries.

The Intergovernmental Panel on Climate Change (IPCC) is another key organization involved in global climate governance. The IPCC provides policymakers with regular scientific assessments of climate change, its impacts, and potential mitigation and adaptation options. These assessments form a scientific basis for climate-related policies and inform decision-makers worldwide. The work conducted by the IPCC is crucial in shaping the understanding of climate change and informing effective policy responses.

In addition to these overarching organizations, there are also specialized agencies that focus on specific aspects of climate governance. The World Meteorological Organization (WMO) plays a vital role in monitoring the Earth's climate system and providing comprehensive data on climate patterns. By doing so, the WMO assists in the development of policies related to climate adaptation and disaster risk reduction.

The United Nations Environment Programme (UNEP) also plays a crucial role in global climate governance. UNEP works towards enhancing environmental sustainability and promotes sustainable development by catalyzing improvement policies and cooperation at both national and international levels. UNEP assists countries in developing environmental programs, including those related to climate change mitigation and adaptation.

Apart from these organizations, regional bodies also have an important role to play in global climate governance. The European Union (EU), for example, is at the forefront of international climate action. Through its policies, legislation, and initiatives, the EU has demonstrated strong leadership in combating climate change. The EU's commitment to reducing greenhouse gas emissions and transitioning to a low-carbon economy has set an example for other regional blocs and countries.

These global climate governance organizations, whether global or regional, play a vital role in coordinating efforts, promoting scientific research, sharing knowledge, and disseminating best practices. They bring together governments, scientists, NGOs, and other stakeholders to foster cooperation, set goals, and review and update climate-related policies. These organizations provide a platform for collaboration and ensure that climate change is approached collectively, with a shared sense of responsibility.

Moreover, global climate governance organizations also contribute to the mobilization of financial resources to support climate-related programs. Through various financial mechanisms established under their purview, such as the Green Climate Fund, these organizations ensure that developing countries receive the necessary support to implement climate adaptation and mitigation projects. This financial assistance is crucial in addressing the unequal burdens of climate change and facilitating a smoother transition towards a low-carbon future.

In conclusion, global climate governance organizations, such as the UNFCCC, IPCC, WMO, UNEP, and regional bodies like the EU, play an indispensable role in addressing the complex and pressing issue of climate change. They facilitate international cooperation, provide scientific assessments, support policy development, and mobilize financial resources. By working collectively, these organizations enable the world to tackle climate change effectively and foster a sustainable and resilient future.

Section 8.3: The Role of Global Climate Governance Organizations

Chapter 9: Engaging Public Opinion and Climate Activism

Public opinion plays a pivotal role in shaping policies and decisions regarding climate change. Engaging public opinion in climate activism is crucial to create awareness, mobilize support, and drive meaningful action towards a sustainable future. This chapter explores various strategies for engaging public opinion and highlights the importance of communication, education, and inclusivity in climate activism.

1: Understanding Public Opinion

1.1 Definition and Significance:

Public opinion refers to the attitudes, beliefs, and perceptions held by the general public on a particular issue. In the context of climate change, public opinion can influence political will and drive action.

1.2 Current State of Public Opinion on Climate Change:

Despite increasing scientific consensus on climate change, public opinion varies across different regions and demographics. While some segments of society are actively engaged and concerned about climate change, others may be indifferent or skeptical.

1.3 Attitude-Behavior Gap:

One challenge in climate activism lies in bridging the gap between pro-environmental attitudes and behaviors. Understanding the factors contributing to this gap is essential for effective public engagement.

2: Strategies for Engaging Public Opinion on Climate Change

2.1 Effective Communication:

Communicating climate science in an accessible and relatable manner is key to engaging public opinion. Using clear language, visual aids, and relatable stories can enhance understanding and relevance for diverse audiences.

2.2 Targeted Messaging:

Tailoring messages to specific demographic groups can increase resonance and engagement. Understanding the values and interests of different segments of the population allows for more effective communication.

2.3 Education and Awareness:

Promoting climate education and awareness at various levels, from schools to community organizations, is crucial in fostering a broad understanding of climate issues. This can empower individuals to take action and contribute to collective efforts.

2.4 Inclusivity and Representation:

Ensuring diversity in climate activism is essential to engage wider public opinion. Including voices from marginalized communities and giving them representation creates a sense of ownership in climate action and helps address social inequalities.

2.5 Collaboration and Partnerships:

Building alliances with diverse stakeholders, including NGOs, businesses, religious groups, and government agencies, can amplify the impact of climate activism. Collaboration allows for diverse perspectives and resources in addressing climate challenges.

3: Tools for Climate Activism and Public Engagement

3.1 Digital Platforms and Social Media:

Leveraging the power of social media and online platforms can exponentially increase reach and engagement. From creating viral campaigns to information dissemination, digital tools enable climate activists to mobilize public opinion globally.

3.2 Public Events and Demonstrations:

Organizing public events, protests, and demonstrations provides a platform to mobilize support and create visibility. It generates public discourse and amplifies the voices of climate activists, influencing public opinion.

3.3 Grassroots Organizing:

Harnessing the power of grassroots movements can instigate social change at the local level. Involving communities directly affected by climate change empowers them to advocate for climate justice and fosters a sense of environmental stewardship.

Engaging public opinion in climate activism is a multifaceted endeavor that requires understanding, strategic communication, education, inclusivity, and collaboration. By employing these strategies and utilizing various tools, climate activists can mobilize public opinion, drive policy changes, and work towards a sustainable and resilient future for all.

Chapter 9: Engaging Public Opinion and Climate Activism

9.1: Enhancing Climate Communication

Climate change is an urgent and complex issue that requires effective communication strategies to raise awareness and engage individuals and communities in mitigating its impact. Enhancing climate communication is crucial in bridging the gap between scientific research and public understanding. This will explore various approaches to improve climate communication, including the use of visual aids, storytelling, and framing techniques.

1. Visual Aids: Visual aids can play a significant role in enhancing climate communication as they appeal to the human visual system, making complex information more accessible and engaging. Graphs, charts, and infographics can be used to illustrate the significance of climate change, present scientific findings, and convey the potential consequences of inaction. Additionally, photographs and satellite images can provide a visual representation of the effects of climate change, such as melting ice caps, deforestation, and extreme weather events. By employing visual aids, climate communicators can effectively convey the urgency and seriousness of the issue, fostering understanding and encouraging action.

2. Storytelling: Effective climate communication often involves telling stories that resonate with people's emotions and experiences. Stories have a unique ability to captivate audiences and elicit empathy, making complex issues more relatable and memorable. Climate communicators can leverage personal narratives that highlight human experiences and connect them to the broader climate change context. Sharing stories of individuals facing the consequences of extreme weather events, farmers adapting to changing agricultural conditions, or communities implementing innovative climate solutions can inspire action and motivate behavior change. By employing storytelling techniques, climate communicators can motivate individuals to become actively involved in addressing climate change.

3. Framing Techniques: Climate communication can also benefit from strategic framing, shaping the presentation and understanding of climate-related information. Framing involves emphasizing specific aspects of an issue to influence how individuals perceive and interpret the information. Climate

communicators can use different frames to appeal to different audiences, addressing their values, beliefs, and concerns. For example, economic framing emphasizes the potential economic benefits of renewable energy and environmentally-friendly initiatives, appealing to business owners and policymakers. Health framing highlights the health risks associated with climate change, targeting healthcare professionals and individuals concerned about wellness. Framing techniques can help engage diverse sectors of society and foster a broad-based understanding of climate change.

In conclusion, enhancing climate communication is vital in mobilizing collective action to address climate change. By utilizing visual aids, storytelling, and framing techniques, climate communicators can effectively convey the urgency, relevance, and potential solutions of climate change to different audiences. With improved communication strategies, we can bridge the gap between scientific research and public understanding and accelerate the transition towards a sustainable and resilient future.

Section 9.1: Enhancing Climate Communication

9.2: Grassroots Movements for Climate Justice

In recent years, there has been a surge in grassroots movements aimed at achieving climate justice. These movements are driven by passionate individuals who are deeply concerned about the impact of climate change and demand immediate action to address it. This will delve into the key aspects of grassroots movements for climate justice, highlighting their significance and the strategies they employ.

One of the defining characteristics of grassroots movements for climate justice is their long-term vision. These movements recognize that climate change is not a standalone issue, but rather interconnected with a wide range of social justice concerns, such as economic inequality, racial discrimination, and indigenous rights. They see climate justice as a framework that encompasses all these issues and seek to address them holistically.

Moreover, grassroots movements for climate justice aim to amplify the voices of marginalized communities who are disproportionately affected by the impacts of climate change. These communities, often from low-income backgrounds or disadvantaged regions, face the brunt of climate-related disasters, such as flooding, extreme heatwaves, and food scarcity. Grassroots movements strive to ensure these voices are heard and integrated into decision-making processes at local, national, and international levels.

To achieve their goals, grassroots movements employ a diverse range of strategies and tactics. Direct action is a prominent feature, with activists engaging in protests, sit-ins, and civil disobedience to bring attention to the urgency of climate-related issues. By disrupting normal operations, these actions aim to garner media coverage and push policymakers to take concrete action.

Social media and online platforms have also become crucial tools for grassroots movements, enabling them to connect and mobilize supporters around the globe. Online campaigns, viral videos, and hashtags are widely used to spread awareness, share stories, and call people to action. This digital landscape has expanded the reach and impact of grassroots movements, helping them tap into a wider audience base and build solidarity among diverse communities.

Grassroots movements for climate justice are further characterized by their decentralized structure and emphasis on decentralized decision-making. Decision-making is often made through collective processes, such as consensus-building and community-driven approaches. This enables a sense of ownership and empowerment among activists, whereby they are mobilized to take action and priorities are collectively determined, strengthening the movement's resiliency and effectiveness.

Furthermore, these movements strive to create alternative systems and solutions that challenge the status quo. They advocate for sustainable practices, renewable energy sources, and climate-friendly policies. By promoting localized, community-based initiatives, grassroots movements not only work towards climate justice but also foster socioeconomic resilience and empowerment at the local level.

In conclusion, grassroots movements for climate justice play a pivotal role in advocating for immediate action to address climate change comprehensively. By recognizing the intersectionality between climate change and various social justice concerns, these movements aim to amplify the voices of marginalized communities and create alternative solutions. Empowered by digital platforms and deploying diverse strategies, grassroots movements strive to bring about systemic change that ensures a just and sustainable future for all.

Section 9.2: Grassroots Movements for Climate Justice

9.3: Institutional Advocacy and Political Change

Institutional advocacy and political change play crucial roles in shaping societies and promoting positive transformations. This aims to provide a comprehensive understanding of this concept, highlighting its importance and impact on various aspects of society.

At its core, institutional advocacy refers to the act of influencing decision-makers within socio-political institutions to drive policy changes and initiate reform. It involves utilizing various strategies and tools to promote a specific agenda, often centered around social justice, equality, human rights, or climate change. This form of advocacy recognizes the significance of engaging with formal structures and mechanisms of power in order to create lasting and impactful change.

One of the key elements of institutional advocacy is its focus on academic, research-based analysis to inform policy debates and decisions. By backing arguments and proposals with solid evidence and data, advocates can better validate their claims and increase the chances of their suggestions being implemented. This integration of knowledge and expertise from various fields helps ensure that policy outcomes are informed and effective.

Moreover, institutional advocacy aims to shift societal attitudes towards certain issues through strategic messaging and storytelling. Recognizing the power of narratives, advocates often employ emotional appeals and personal anecdotes to convey the urgency and importance of their cause. By humanizing complex policy problems and connecting them to real-life experiences, advocates are able to garner public support, which is vital for political change.

The impact of institutional advocacy in driving political change can be seen in various contexts. For instance, it has played a pivotal role in influencing legislation addressing gender equality and LGBTQ+ rights. By engaging key decision-makers within political institutions, proponents of these movements have been able to promote inclusive policies that protect and advance the rights of marginalized communities.

Furthermore, institutional advocacy has been critical in advancing environmental causes such as climate change mitigation and sustainability. By mobilizing activists, engaging with policymakers, and utilizing scientific research, advocates have been able to shape governmental initiatives and international agreements aimed at combating climate change. This form of advocacy has prompted governments to adopt renewable energy targets, commit to emission reductions, and invest in sustainable development.

In addition, institutional advocacy has the potential to bring about changes in healthcare policies, education reform, criminal justice, and economic development. By effectively leveraging their resources and networks, advocates can influence the priorities of political institutions, redirect resources towards neglected sectors, and instigate systemic transformations.

However, it is important to acknowledge the challenges and limitations that come with institutional advocacy. As decision-making processes within political systems can be slow and complex, advocates may encounter resistance, bureaucracy, and competing interests. Overcoming these barriers requires navigating power dynamics, building strong coalitions, and adopting long-term strategies.

In conclusion, institutional advocacy and political change play crucial roles in shaping societal progress. By engaging with decision-makers and utilizing research-based analysis, advocates can contribute to policy reform and drive positive transformations. Through their strategic messaging and ability to mobilize public support, they have the power to shift societal attitudes and advance social justice causes. While challenges may arise, the potential for institutional advocacy to drive meaningful change highlights its importance within the broader context of democratic societies.

Section 9.3: Institutional Advocacy and Political Change

Chapter 10: Climate Literacy in School Curricula and Educational Institutions

In recent years, the need for climate literacy has become increasingly evident as the impacts of climate change continue to manifest. Educational institutions have a crucial role to play in equipping students with the knowledge and skills necessary for understanding and taking action on climate change. This chapter aims to explore the incorporation of climate literacy in school curricula and highlight the strategies and challenges faced by educational institutions in this process.

The Importance of Climate Literacy:

Climate change represents one of the most pressing challenges of our time. It affects not only the environment but also social, economic, and political systems. Climate literacy is crucial for individuals to understand and address these interconnected challenges effectively. By incorporating climate literacy into school curricula, educational institutions can empower students to become informed global citizens and agents of change.

Integrating Climate Literacy into School Curricula:

Several countries have taken significant steps to integrate climate literacy into their school curricula. This includes the inclusion of climate change topics in science, geography, and social studies classes. Curricula focus on fundamental concepts such as the greenhouse effect, the carbon cycle, and the impacts of climate change on ecosystems, societies, and economies.

In addition to subject-based learning, schools also emphasize interdisciplinary approaches to climate literacy. They encourage students to explore climate change through various lenses, including ethics, economics, politics, and culture. This interdisciplinary approach fosters the understanding that climate change is a multi-faceted issue that requires diverse perspectives for effective solutions.

Strategies for Effective Climate Literacy Education:

Climate literacy education can be enhanced through several strategies. First, schools should prioritize hands-on learning experiences, such as field trips and outdoor activities, to foster a direct connection between students and the environment. This helps students witness the impacts of climate change firsthand and develops their understanding of environmental systems.

Furthermore, interactive technology and digital platforms can play a vital role in climate literacy education. Virtual simulations, educational games, and online resources provide engaging and interactive ways for students to learn about climate change. These tools appeal to modern students and facilitate active engagement with the subject matter.

Challenges Faced:

However, despite the efforts made, there are several challenges confronting schools in incorporating climate literacy into their curricula. One major challenge is the lack of trained teachers who possess a strong understanding of climate science. Many teachers feel ill-prepared to teach climate change-related topics due to limited training, resources, and confidence. Addressing this challenge requires investment in professional development programs and increased support for teachers.

Another challenge lies in the politicization of climate change education. In certain regions, curriculum decisions are influenced by political agendas, leading to the exclusion of accurate and evidence-based information on climate change. Overcoming this challenge requires ongoing advocacy and dialogue to ensure climate literacy is not compromised by political interference.

Climate literacy should be an essential component of school curricula to equip students with the knowledge and skills necessary for addressing climate change. By incorporating climate literacy into their curricula, educational institutions can inspire future generations to become informed, engaged, and proactive global citizens. Despite the challenges faced, continued efforts are essential to ensure climate literacy becomes a foundational aspect of education worldwide.

Chapter 10: Climate Literacy in School Curricula and Educational Institutions

10.1: Assessing Current Climate Education Standards

Assessing current climate education standards is essential for ensuring that our future generations are equipped with the knowledge and skills necessary to tackle the challenges of climate change. In this section, we will delve into the assessment of existing climate education standards, evaluating their efficacy, scope, and impact on educators and students alike. By analyzing these standards, we can identify the strengths and weaknesses of current educational practices, as well as potential areas for improvement.

Assessment Methods:

Assessing climate education standards requires comprehensive and rigorous evaluation techniques. One commonly used method is the content analysis of standards documents. This involves a systematic examination of the stated objectives, content, and learning outcomes outlined in the standards. By examining the coherence and depth of the content, we can assess the comprehensiveness of the standards in addressing climate change.

Another effective assessment method is conducting surveys and interviews with educators and students. This provides valuable insights into the implementation and usefulness of the standards in classroom settings. Feedback from educators can help identify challenges faced during implementation, while student perceptions reveal the level of engagement and understanding achieved through the standards.

Furthermore, reviewing research literature on climate education can provide valuable insights. Studying the impact of climate education standards on student learning outcomes can help determine their effectiveness. Additionally, comparing standards across different regions or countries can highlight best practices and areas where improvements can be made.

Evaluating Efficacy:

Assessing the efficacy of current climate education standards involves measuring their impact on knowledge and behavioral changes. Standardized tests can be used to assess the acquisition of climate-related knowledge among students. These tests should align with the objectives and content of the standards, measuring proficiency in understanding climate science, impacts, mitigation strategies, and adaptation measures.

Behavioral changes regarding climate action can also be assessed through surveys and observations. This includes evaluating changes in attitudes, beliefs, and actions related to climate change. Educators can observe students' behaviors, such as energy-saving practices or participation in climate-oriented initiatives, to gauge the influence of education on behavior.

Scope and Depth:

Assessing the scope and depth of climate education standards requires a thorough analysis of the curriculum frameworks and included topics. Topics should be structured to cover fundamental concepts of climate science, as well as interdisciplinary relationships with other subjects, like geography, biology, and social sciences. Assessing the alignment between standards and other related disciplines ensures a holistic understanding of climate change, addressing its scientific, social, and economic dimensions.

Assessing the depth of standards involves evaluating the level of complexity and critical thinking skills demanded from students. Climate education should not focus solely on imparting facts but should also encourage students to analyze data, evaluate evidence, and develop logical reasoning skills. By assessing the depth of standards, educators can gauge whether students are adequately equipped to understand and participate in discussions surrounding complex climate issues.

Assessing current climate education standards plays a crucial role in improving the quality and effectiveness of climate education. Through content analysis, surveys, interviews, and the evaluation of efficacy, scope, and depth, educators and policymakers can identify areas for improvement, ensure a comprehensive understanding of climate change, and promote meaningful behavior change. By consistently assessing climate education standards, we can foster a more informed and prepared generation ready to face the challenges of a changing climate.

Section 10.1: Assessing Current Climate Education Standards

10.2: Best Practices and Innovations in Climate Teaching

The study and understanding of climate change have become increasingly important in recent years. As educators, it is crucial for us to keep up with the latest developments in this field and incorporate the best practices and innovations in climate teaching into our pedagogical approaches. In this section, we will delve into some of the most effective strategies and techniques that can be used to engage students and promote their understanding of climate change.

One of the best practices that has emerged in climate teaching is the incorporation of real-world evidence and case studies into the curriculum. Climate change is not an abstract concept, and it is important for students to see concrete examples of its impact around the world. By using case studies such as the melting of polar ice caps or the devastation caused by extreme weather events, educators can help students grasp the gravity of the situation and motivate them to take action.

Another innovative approach to climate teaching is through the use of interdisciplinary projects. Climate change is a complex issue that cannot be understood through a single lens. By engaging students in projects that integrate multiple disciplines such as science, geography, economics, and social studies, educators can provide a holistic understanding of the problem. For example, students can examine the environmental and economic impacts of renewable energy initiatives or analyze the social and political factors that contribute to climate policy decisions.

Technology also plays a significant role in enhancing climate teaching. With the rise of digital tools and online resources, educators can incorporate interactive simulations, virtual field trips, and data analysis activities into their lessons. These technological advancements not only make the learning experience more engaging for students but also allow them to explore the complexities of climate change in a hands-on manner. For example, students can use online platforms to track and analyze real-time weather data, or they can participate in virtual simulations that illustrate the impacts of rising sea levels.

In addition to these strategies, it is important to encourage critical thinking and problem-solving skills in climate teaching. Climate change is an inherently complex issue, and it requires students to analyze data, evaluate sources, and make informed decisions. By incorporating activities that promote these skills, such as debates, case studies, and research projects, educators can empower students to become active participants in the fight against climate change.

Lastly, it is crucial for educators to cultivate a sense of hope and empowerment among students. The topic of climate change can often feel overwhelming and hopeless, leading to a sense of apathy or resignation. It is important to counteract these feelings by highlighting success stories and showcasing the impact of individual and collective actions. By emphasizing the solutions and opportunities for positive change, educators can inspire and motivate students to become agents of change.

In conclusion, the best practices and innovations in climate teaching involve the incorporation of real-world evidence, interdisciplinary projects, technology, the promotion of critical thinking skills, and the cultivation of hope. By implementing these strategies, educators can ensure that students develop a comprehensive understanding of climate change and are equipped with the knowledge and skills necessary to tackle this critical issue. Through effective climate teaching, we can empower the next generation to shape a sustainable future.

Section 10.2: Best Practices and Innovations in Climate Teaching

10.3: Overcoming Barriers to Implement Climate Education

In recent years, there has been increasing recognition of the need for climate education to be incorporated into the curriculum in schools and higher education institutions worldwide. The urgency of addressing climate change and reducing carbon emissions has made it imperative that individuals, communities, and governments are well-informed about the impacts of climate change and their role in creating a sustainable future. However, the process of implementing climate education faces several barriers that need to be overcome in order to ensure its successful integration.

One major barrier to implementing climate education is the lack of qualified teachers who can effectively deliver the curriculum. Climate change is a complex and multi-faceted topic that requires knowledgeable individuals who can effectively communicate the scientific concepts and societal implications in a pedagogically sound manner. However, many teachers lack the necessary training and expertise in climate science to confidently teach the subject. Overcoming this barrier requires investing in professional development and training programs to equip teachers with the knowledge and skills necessary to effectively teach climate education.

Another barrier is the resistance from educational institutions and policymakers to include climate education in the curriculum. This resistance can stem from a variety of reasons, including political ideologies, concerns about the cost of implementation, and a belief that climate education is not a priority. Overcoming this barrier necessitates a concerted effort to advocate for climate education and demonstrate its relevance to students' lives and future careers. Building alliances with educational leaders, policymakers, and stakeholders is crucial in overcoming resistance to ensure the adoption and implementation of climate education.

Limited access to resources and materials on climate change is yet another barrier to implementing effective climate education. Exposing students to accurate and up-to-date information helps them develop an informed and critical understanding of the subject matter. However, many schools and educators face challenges in accessing reliable and age-appropriate resources that are specifically designed for teaching climate change. Addressing this barrier requires the development and dissemination of high-quality, accessible, and engaging educational resources that cater to different levels and subjects.

In addition to these systemic barriers, the lack of widespread public understanding and acceptance of climate change poses challenges in implementing climate education. Climate change denial or skepticism remains prevalent in some communities, which can hinder efforts to teach about climate change objectively and scientifically. Overcoming this barrier entails fostering a culture of open dialogue, critical thinking, and evidence-based decision-making. Engaging with stakeholders, including community members, parents, and local leaders, is crucial to address misconceptions and increase public understanding of climate change, its impacts, and the importance of climate education.

Furthermore, the integration of climate education across subject areas and grade levels is often limited, with climate-related topics being siloed within a specific subject or grade level. This compartmentalization restricts students' comprehensive understanding of climate change and its interdisciplinary aspects. To address this barrier, climate education should be integrated across the curriculum, incorporating it into subjects such as science, geography, social studies, and even literature and the arts. Such an approach ensures that students receive a holistic and interconnected understanding of climate change and its implications.

In conclusion, implementing climate education requires proactive efforts to overcome a range of barriers. By investing in teacher training and professional development, advocating for its inclusion in the curriculum, providing access to reliable resources, addressing public understanding, and integrating climate education across subjects, we can strive to ensure that students receive an education that equips them with the knowledge, skills, and values necessary to navigate the complex challenges posed by climate change.

Section 10.3: Overcoming Barriers to Implement Climate Education

In conclusion, the importance of empowering a climate-literate generation cannot be overstated. As we face critical environmental challenges in the 21st century, it is imperative that we equip our youth with the knowledge and skills necessary to understand, address, and mitigate the impacts of climate change.

Empowering the younger generation with climate literacy starts with education. It is through comprehensive and interdisciplinary education that young minds can grasp the complexity of climate change and its implications for our planet. By integrating climate science into school curricula, we can ensure that students are well-equipped with the basic knowledge and understanding of climate systems, greenhouse gases, and the various factors contributing to global warming.

Beyond classroom education, it is essential that we foster critical thinking and engagement amongst our youth. Empowering young individuals to think critically about climate change not only enables them to make informed decisions but also encourages them to question and challenge existing practices that perpetuate environmental degradation. By instilling a sense of agency, we can inspire young people to become advocates for change and empower them to take action on a personal, community, and global scale.

Furthermore, creating opportunities for youth to engage in firsthand experiences can significantly enhance their climate literacy. This can be achieved through field trips, outdoor activities, and hands-on projects that allow students to witness the effects of climate change firsthand. By connecting young individuals with nature and the environment, we can foster a deep understanding and appreciation for the fragile balance of our ecosystems, encouraging them to actively participate in conservation efforts.

In addition to education and engagement, technology and media play a crucial role in empowering a climate-literate generation. Through online platforms, social media, and digital communication tools, we can amplify the voices of young activists, educators, and scientists who are working tirelessly to combat climate change. By harnessing the power of technology, we can create networks and communities that foster knowledge sharing, collaboration, and innovation, enabling young individuals to stay informed and engaged in the fight against climate change.

Finally, it is essential to involve young people in decision-making processes and provide them with opportunities to contribute their perspectives and ideas. Empowering youth as stakeholders not only ensures that their voices are heard but also recognizes the value of their unique insights and solutions. By involving young people in policymaking, implementing youth-led initiatives, and fostering intergenerational collaboration, we can harness the creativity and passion of our youth to drive meaningful change and create a more sustainable future.

In conclusion, empowering a climate-literate generation is crucial for the future of our planet. By equipping young individuals with climate knowledge, fostering critical thinking and engagement, providing firsthand experiences, utilizing technology and media, and involving young people in decision-making processes, we can empower the next generation to become lifelong stewards of the environment. It is through their collective efforts, innovation, and determination that we can confront the challenges posed by climate change and create a more sustainable and resilient planet for generations to come.

Conclusion: Empowering a Climate-Literate Generation

In conclusion, empowering a climate-literate generation is of utmost importance in addressing the pressing issue of climate change. The writing has delved into the concept of climate literacy, highlighting how it goes beyond mere knowledge and encompasses a deep understanding of the complexities of climate change. It has emphasized the need for individuals to be equipped with the skills to critically analyze information, make informed decisions, and actively engage in climate action.

The writing has also provided a comprehensive overview of the benefits that come with climate literacy. It has elucidated how climate-literate individuals are more likely to adopt sustainable and responsible behaviors, reducing their carbon footprint and contributing to positive environmental change. Moreover, climate literacy enables individuals to grasp the global consequences of climate change and fosters a sense of solidarity and interconnectedness with the world.

Furthermore, the writing has shed light on the role of education in cultivating climate literacy. It has emphasized the need for educational institutions to incorporate climate change into their curricula, providing students with the knowledge, skills, and tools to understand and tackle this critical issue. Additionally, the writing has highlighted the significance of interdisciplinary learning, recognizing that climate change encompasses multiple fields of study.

The writing has also explored some effective strategies for empowering a climate-literate generation. It has discussed the importance of hands-on learning experiences, such as field trips and outdoor activities, in enabling students to witness the real-life impacts of climate change. It also emphasizes the role of technological advancements in enhancing climate literacy, offering interactive tools and platforms that foster engagement and understanding.

Overall, the writing has successfully portrayed the significance of empowering a climate-literate generation. By equipping individuals with the necessary knowledge, skills, and mindset, we can work towards a more sustainable and resilient future. It is through climate literacy that we can foster a generation that actively contributes to climate action, making informed decisions, and advocating for positive change. Therefore, investing in climate literacy today is an investment in the tomorrow we want to secure for ourselves and future generations.

86

www.ingramcontent.com/pod-product-compliance
Lightning Source LLC
Chambersburg PA
CBHW031210160726
47992CB00006B/2667